An Introduction to S

To my wife, Marlena, with all my love.

An Introduction to Social Housing

Second edition

PAUL REEVES MA(CANTAB) MCIH PGCE(HE) MCMI

ELSEVIER
BUTTERWORTH
HEINEMANN

AMSTERDAM BOSTON HEIDELBERG LONDON NEW YORK OXFORD PARIS
SAN DIEGO SAN FRANCISCO SINGAPORE SYDNEY TOKYO

Elsevier Butterworth-Heinemann
Linacre House, Jordan Hill, Oxford OX2 8DP
30 Corporate Drive, Burlington, MA 01803

First published in 1996 by Arnold

Second edition 2005

ISBN 0 7506 63936

British Library Cataloguing in Publication Data
A catalogue record for this book is available from the British Library

For information on all Butterworth-Heinemann publications
visit our website at http://bookselsevier.com

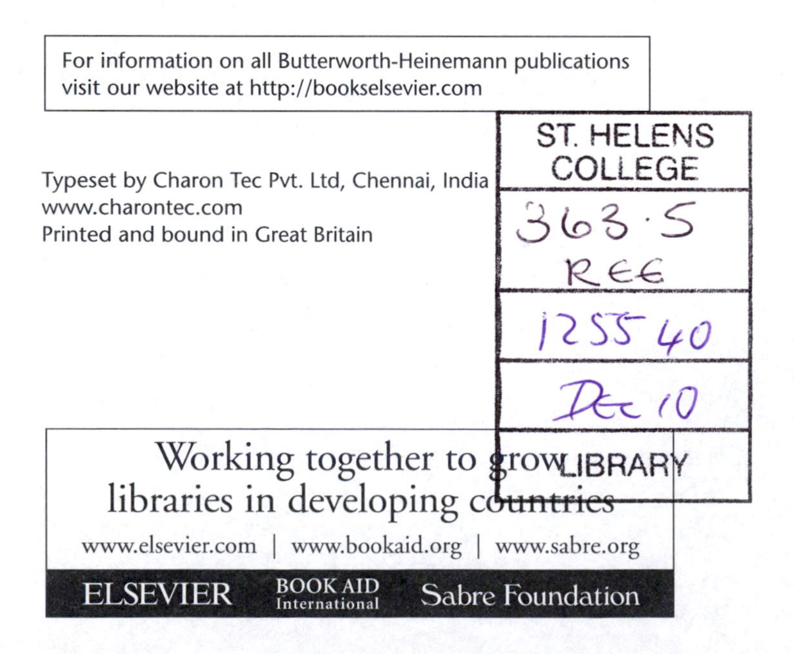

Typeset by Charon Tec Pvt. Ltd, Chennai, India
www.charontec.com
Printed and bound in Great Britain

Contents

Acknowledgements

I could not have written the second edition of *An Introduction to Social Housing* without having worked for and lectured at a number of institutions. Each workplace and lectureship contributed a great deal to my knowledge and understanding of the subject, and I am indebted to them all.

I would like to thank all relevant staff past and present at the following local authorities where I worked between 1982 and 2004:

- Cambridge City Council
- London Borough of Southwark
- London Borough of Greenwich
- London Borough of Richmond on Thames
- London Borough of Islington
- Association of London Government

Personnel in other organisations I have worked for, and held Board positions at, have also been a source of invaluable information and support, including:

- Fordham Research
- Tenant Participation Advisory Service
- Swale Housing Association (now AMICUS)
- Bush Housing Association
- Garden City Homes Housing Association

Fellow lecturers and students at the following establishments should be thanked for all they have done in responding to lectures and for giving constructive criticism:

- Hammersmith and West London College
- University of Westminster
- University of Greenwich
- Anglia Polytechnic University

Above all, I could not have contemplated this work without the loving support and patience of my wife, Marlena, to whom this book is dedicated.

Paul Reeves
Surrey

List of figures and tables

Introduction

The provision and management of social housing for those who are unable to access the housing market is essential to the maintenance of the fabric of society. Roughly 20 per cent of households in this country rely upon some form of subsidised housing provided by local authorities and housing associations, and many who would otherwise be homeless are housed in private sector accommodation procured by state and voluntary agencies. Yet others rely on housing benefits provided through tax receipts to help them afford the homes they rent. The social housing industry is vast and still growing, with an annual growth in the number of housing associations and management bodies, and is changing to adapt to new political and economic forces. There are very few countries in the world where some form of subsidised housing does not exist, and the total number of social homes is likely to grow worldwide, as are the challenges of the sector.

This book is aimed at students on housing and related courses, ranging from HNC level through professional qualifications to housing and related degrees. It covers the main themes in the field, giving a broad overview as well as detailed case studies exemplifying housing policy and practice in a mainly UK and specifically English context, although drawing on best practice elsewhere, where alternative non-UK approaches are thought to be appropriate to the issues explored. It is also meant for housing professionals at all levels who desire a broader overview of the subject than may be gained through working in a specific part of, or function in, the sector. The overall aim is to enhance understanding of the key themes in social housing, so as to increase the effectiveness of housing management, and so the quality of life of those who depend on the sector for a roof. It is hoped that it may motivate some to consider or implement positive changes, which will move housing policy and practice forward, in an uncertain and developing social and economic environment where change is a constant.

Social housing is a phrase which has only really gained currency over the last fifteen years or so, and it is not without its critics, both as an expression and as a concept. In this book, it is taken to mean housing provided by local authorities and housing associations (sometimes known as Registered Social Landlords, although there is not an exact equivalence), and extended to cover housing managed by these bodies, regardless of ownership. The key feature which defines the essence of the products and services provided by all social housing providers is that these activities and products are non-market, in that they cannot be obtained by bidding with cash or other financial resources in competition, and that the products are allocated principally on the basis of housing need rather than effective demand, although there has been some blurring of the boundary with the increased prominence of low-cost home ownership in general and shared ownership housing in particular.

Social housing bodies can be direct providers of housing – such as when a housing association develops and manages property, or enablers – where a body helps its clients or customers to find housing through another agency, for example, where a local authority (that is, a council) makes a cash grant to a housing association to build housing, or works through the planning system to enable land for social housing development. Another example of enabling is where a council contracts out the management of its homes to another party. The largest housing enabler is the government, which makes cash help available to back council housing management and maintenance activities,

t the child's, but held in trust for
agency which helps fund housing
t, the Housing Corporation, is always
t at the end of the day, it will decide
ttle will be given out, or recycled, as it
e the guardian of the public purse, and
needs of local authorities against those of
ions, such as national security, environ-
on and running its bureaucracy. Analogy
omplete accuracy, but can often be usefully
make sense of what at first appears to be
almost poetic dialogue.

ent is the next great theme of the book. It is
ilding flats and houses for rent and for sale which
ow and in the future, who are on lower incomes
annot easily compete in the marketplace, or who are
f the race entirely, can afford to live in, which is suit-
for their needs, and which meets both their aspirations,
east in part, and those of the provider. It links clearly to
e theme of supply and demand: how much, when and
where, are perennial questions raised and not always satisfactorily answered. At times of very great demand, where the state has perceived there to be a 'housing crisis', such as after the First and Second World Wars, the establishment has busied itself in rushing up new towns, enabling the expansion of existing ones, and encouraging the construction of high-density estates and dwelling-types, including tower blocks and the like, to house the people: and at other times, it has encouraged the demolition of homes deemed to be surplus to requirement. Changes in the law, such as the *1925 Building Societies Act*, which made it much easier to get mortgages to buy homes over twenty-five or thirty years, and the consequent growth of lending institutions and products to help relieve individuals of their money over the long term, prompted a massive development of suburbs around London and other major cities from the 1920s onwards. As a result, three-bed semi land has become an enduring feature of the built environment, as much as high levels of owner-occupation as the preferred tenure has become a prominent feature – and one of the most discussed aspect of – the social landscape of this country.

and gives permission to councils to raise money to do major regeneration or improvement schemes. The Housing Corporation, the housing agency responsible to Parliament for part funding housing associations through the social housing grant regime, is another key enabler, as are private finance institutions. It is important to understand the nature and development of providers and enablers to gain a full awareness of the scope and direction of social housing, against a backdrop of macro-economic imperatives.

The historical trend since the late twentieth century has been for councils to relinquish their development and in many cases their management role in favour of enabling via housing associations and even private companies. It was not always thus: from the end of the First World War to the early 1980s, councils were indisputably the key providers of social housing in all the senses described, and a combination of political and economic pressures led to the weakening of the municipal provider role. The key reason for this diminution is related to changing government fiscal and monetary policies, especially a concern to control inflation through public expenditure constraint, on the assumption that the control of public sector finance is probably the easiest and most acceptable way to regulate the amount of money in circulation, rather than imposing credit controls on consumers or by relying heavily upon the crude lever of interest rate adjustments. There is little reason, on the evidence of the past thirty years, to believe that there will be a reversal in the move away from municipal housing provision and management, and it is highly likely that in twenty years hence, the expressed wish of the 1987 Conservative Housing Minister, William Waldegrave, that 'there should not be much of it' (referring to council housing) – either in the ownership or the management sense – will probably be granted. Predictions in this field are always dangerous, but the trend to date is quite clear, and students of housing and practitioners need to be adaptive to changing circumstances, as in any other profession.

The book will deal with the UK social housing situation, but national housing policy, if there can be said to be one, is being increasingly influenced by European law and institutions, as well as by concepts from further afield. Many of the problems and challenges faced by housing managers in this country differ little in type and degree to those elsewhere, and policies and strategies to devise reasoned reactions, measures for containment or even solutions may frequently be found outside the UK or Europe, albeit with modification as a result of cultural, political, social and economic state and inter-state differences. This justifies examining case studies of housing policy and practice elsewhere in illustrating possible ways forward. A good example is given in the choice-based lettings system devised in Delft, Holland, in the mid-1990s, allowing housing applicants to bid for properties according to both their relative needs and wants, which has been successfully applied in the UK in both low and high housing demand contexts. Another is in the development of ecologically and environmentally sustainable housing, pioneered outside the UK, mainly in Scandinavia, which has provided a good role model for development here. And the development of site-and-service plots in the favelas of Rio de Janeiro to enable self-builders to construct viable homes without recourse to massive state subsidy or reliance on the national or local state, providing low cost housing for rent or for sale, is arguably a reasonable model for assisting would-be home owners to realise their ambitions on a low capital input basis here. There are other models, such as an armed housing police presence in many of the USA's welfare housing areas, which should not be lightly replicated in the UK, but the issues surrounding the presence and control of anti-social behaviour are often similar, and it is useful to consider the variety of responses in various cultural situations, if only for the sake of comparison and contrast.

This book will deal with housing in a people-centred manner, looking at enduring themes in housing provision and management, rather than as a handbook or list of current housing study topics, such as the law, housing finance, development, planning, allocations, lettings, and housing associations. It is impossible for a modern book on housing to cover these details in an up-to-date manner, as the legislational and practice context is always changing, and going to relevant websites (for example, the ODPM's or Housing Corporation's) is probably a much better way of getting the up-to-date

facts than readi
ably around
housing ar
These th
basic
live

The
does n
househo
rights and r
to and from th
munity. If the ho
able to without too
tics change, there shou
by moving or by changing
home they live in. These unive
in principle over generations, a
tom and legislation merely provide
will stress the universal and little-changi
realities which drive social housing.

The 'big picture' themes of social housing are
lying housing supply and demand, housing ma
the finance of housing and the political and econom
ities which drives it, equalities and diversity issues w
run through the entire field, and ways of developing ne
housing.

By way of introduction, the study of housing supply and demand demands that we look at the variation in wealth economic performance spatially. In some areas of the country, there is oversupply of housing in relation to demand, and in others, there is a supply shortage. This reality is underpinned by differential economic performance – related largely to the growth and decline of key industries and economic sectors. An example is given by the growth of the financial and related sectors in London and the South East, and the consequent relative skilled labour shortages which have helped generate higher average wage levels through

This is the stuff of development, and the rest – the technical bits including land assembly, planning policy and practice, construction options and technology, the design of sustainable communities if this is ever possible, and the improvement of existing estate layouts and dwellings to address social as well as structural issues – falls into place once the big picture is defined.

The reality of housing studies is that all of these themes are linked, and centred on the needs of households and the constraints imposed by public policy choices and pressures, and it is the links which will be stressed time and time again in this book, to help the reader get a handle on the big picture, which will perhaps motivate them to change the housing system for the better, or at least to understand where they fit in the housing provision and management system.

1 Supply and demand

Housing is not free, and can be very expensive for the customer to obtain and to live in. This may seem an obvious statement, but why isn't it free, or at least cheaper? It is a necessity, but so is air, and that is free. Water is also essential for survival, but does not cost very much to the consumer. The price of housing to the consumer, in both the rented and owner-occupied sectors, and certainly in the social housing area, varies very considerably by area and type, and the underlying reason for this is economics. The amount of housing available compared to demand (and need) for it varies across the country and the world, and over time as well, and it is this relative variation which helps to determine the price of the good, as for any other, at least in the private sector, but increasingly in the social housing sector as well. This price variation is also the major stimulus to social housing programmes, and helps to explain their variation in size – and the need for them – over space and time. The economics of supply and demand also influences rent levels in the social rented sector directly – those familiar with rent restructuring will know that one of the components helping to determine target rent levels is the market value of property, as well as average manual earnings levels, which addresses the element of effective demand.

What has supply and demand got to do with housing?

It is assumed for the purposes of this chapter that students have a basic familiarity with the economic concept of supply and demand: a detailed treatise on the subject is outside the scope of this book. However, fundamentally, the price of most goods is determined by availability and the level of effective demand, where effective demand means demand from people able to pay for the product.

The so-called laws of supply and demand can be summarised as follows. Scarcer things that people want are more expensive than less scarce things, given a similar level of cash-backed demand for both, and the price of things varies with variations in levels of supply and demand. If demand rises and supply remains the same, prices tend to rise. If supply rises, and demand remains the same, then prices tend to fall. There are many number of variants of this law, and much depends on whether one product can be substituted for another, or whether there is no substitute (yet), and on the degree to which people really want the item, and, of course, on fashion, but the basic principle is still recognised as correct by most economists, and has its origins in the work of the eighteenth-century economist Adam Smith, in his influential book, *The Wealth of Nations*.

The concept of price being determined largely by supply and demand is well and relatively simply illustrated in the sphere of housing. A subject which comes up with boring regularity in saloon bars and golf clubs all over South East England, at least in the 1980s to the present, is the way that house prices have risen, and how much money people have notionally made from a sustained rise in property prices over most of the past three decades, with notable blips – but why has the price of owner-occupied housing risen so fast in this region of the UK?

The reason lies in the relative growth of the South-East UK economy, especially in the tertiary (e.g. service, retail, office) and quaternary sector (e.g. finance, information and communications technology, insurance and the like), and a relative shortage of appropriately qualified and skilled personnel able to fill the growing vacancies, leading to significant wage-rises and increases in personal wealth, notably in parts of the finance trading sector. This is a generalisation, and masks important variations in unemployment levels and sub-sectoral and sub-regional wage differentials, but nonetheless helps to explain the relatively high average wages apparent in the South East. This has led to the massive growth of effective demand for owner-occupied housing, again with significant sub-regional variations. Supply has in no way kept pace with demand, for a number of reasons, partly due to developers keeping a lid on production to maintain price levels and therefore profits, the high price of development land, restricting the type of housing which can be made available, planning policy restraint such as the Green Belt policy, weakened in the early twenty-first century, the planning policies of some local authorities, NIMBYism (the 'not in my back yard' attitude of some householders to the possibility of adjacent development) and the more extreme BANANA syndrome in some communities (build absolutely nothing anytime nor anywhere is one interpretation of the acronym), the influence of the environmental lobby and rising river levels threatening floodplain development, especially in parts of the Thames Valley.

The reader is referred to the influential government *Barker Report*, published in 2004 by HM Treasury, which indicated that an additional 120 000 private sector houses would have to be built annually to 2015 in addition to current planning targets in order to bring the price of housing in line with the European Union average (Barker, 2004, p. 11). Crudely, this approach involved calculating the extra supply which would bring prices down to cope with predicted excess demand. The author also deduced that between around 17 000 and 23 000 social homes in addition to existing targets would be required to cater for households who could not exercise effective demand, even if the additional private sector house building targets were met (Barker, 2004, p. 11).

So with demand high and growing, and supply not keeping pace at all, house prices have continued their seemingly inexorable rise, cooled only occasionally by interest rate hikes and recessions reducing effective demand levels.

The effect of this has been to price many households out of the market, especially those not already owner-occupiers and in the early stages of formation, or migrating from one region to another in search of better work prospects, or those with below average incomes.

Rental levels in some areas of very high effective demand for owner occupation and hence high house prices have also risen significantly, in response to the growing surplus unmet effective demand for house purchase, themselves varying with the changing relationship between owner occupation supply and demand.

The effect of such hikes has been to exclude very many households from the possibility of owner-occupation, at least in the short term, and from private rental of suitably sized and located accommodation as well. In such cases, demand is ineffective, and the prospect of unmet need looms. There is therefore a significant and rising level of unmet housing need, which is related to but not the same as the economic concept of demand, in the South East of England, as well as in other parts of the UK. The EU in general displays similar supply–demand characteristics. In crude terms, many households currently cannot afford to become owner-occupiers or private renters in the private sector; hence it is said that there is a shortage of affordable housing.

What is affordable housing?

Affordable housing is, out of its proper context, an odd phrase. After all, most housing is affordable by someone – proved by the fact that it is occupied, or by estate agencies taking properties in for sale, and if there were properties which were genuinely not affordable, they would be empty, but there are few empty houses around in the South East which are inhabitable, unless they are simply temporarily unoccupied and awaiting sale.

Official figures published in 2003 showed that in 2001, roughly 3% of housing in all tenures was vacant – nearly 700 000 dwellings, compared to 800 000 in 1996. The social sector was more likely to have empty property than

Table 1.1 Vacant dwellings by ownership 2001 – percentages of total stock, England (total stock: 683 000)

Ownership	Percentage of vacant stock (rounded)
Registered social Landlord	9
Formerly owner-occupied	47
Formerly private rented	28
Local authority	18

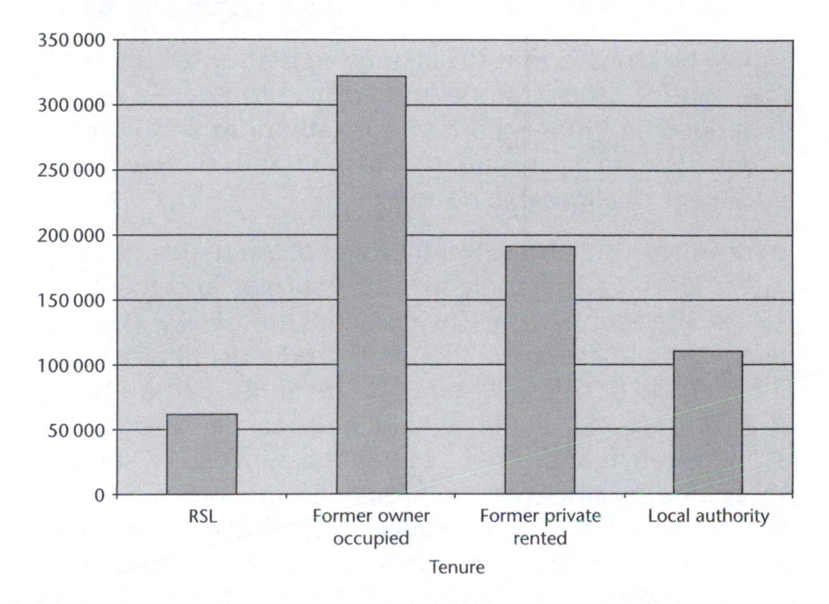

Fig. 1.1 Empty homes in England (Source: English House Conditions Survey 2001 (2003) p. 34).

the private sector (4% vacant compared to 3%) (English Housing Conditions Survey, 2001).

So 'affordable housing' in this context really means the number of houses and flats which households who can't afford to compete in the marketplace need. The need can be calculated by deducting the number of households who have not yet satisfied their housing needs who can do so in the marketplace at a chosen point in time from the total number of households who have not yet satisfied their housing needs. In terms of homes to be built, the output which would satisfy the requirement includes but is not equivalent to the social housing programme required, and is far larger than actual social housing development programmes over the last few decades. It also includes properties which could be built or converted for owner occupation at prices below prevailing market rates (for example shared-ownership housing) or private rented property let at levels below those generally to be found through private lettings agencies. Town and country planners almost always talk about affordable rather than social housing when estimating land requirements for housing provision, and planning legislation is tenure-neutral and similarly refers.

It must be stressed that social housing is a subset of affordable housing, and is properly used only to refer to homes developed and managed by local authorities and housing associations for rent, and should not be regarded as an equivalent to affordable housing.

In summary, then, there is a shortage of affordable housing in the South East of England, and a surplus in some other parts of the country. The Government's *Sustainable Communities Plan*, published in 2003, indicated a need for around 200 000 additional new homes in the South East by 2016, of which around 31 000 would have to be 'affordable', partly to retain and attract essential 'key workers' such as police and health workers to service the region's social infrastructure (The Communities Plan, 2003). Since labour and households are not perfectly mobile, the problems associated with these imbalances are likely to remain, and need to be addressed if the economy of the country as a whole and of its regions is not to continue to be severely damaged.

Demand and need

We have already discussed the distinction between demand and need, but it is necessary to break 'need' down into its constituent categories, because housing need is not standard across cultures and other societal groups. In housing speak, need has been separated into general and special needs categories, although the latter term is no longer as fashionable as it once was. This is because the term 'special' can be stigmatising, indicating without any further analysis that the household referred to is somehow less able or capable of finding and maintaining their home than other groups, due to some defect of character, either on a personal or group basis. However, there has been no particularly acceptable substitution for this term: 'supported housing' could also be said to have similar connotations, although it tends only to imply that the household requires services not needed by more standard cases to look after their home and themselves.

General needs housing is housing provided for households who do not require extra housing or related support, apart, in some cases, from financial assistance towards the rent. General needs customers are provided with homes which vary mainly in response to household characteristics, for example, the number of people and ages of members. The larger proportion of council and housing association stock is provided for general needs groups. The general needs group is very diverse, culturally, ethnically and in terms of sex of head of household and family size, and the level of unmet need amongst its constituent part varies widely, depending mainly on relative income levels, as well as with the degree of discrimination which some groups face in obtaining housing through alternative routes.

Special needs is a term which can be used to cover a variety of households united only insofar as they have some characteristic, physical or mental, which means that they require a higher level of housing or care support than general needs groups. Frequently, the housing provided has to be designed or adapted especially to cater for the specific need or range of needs thought to be significant by the provider, sometimes in consultation with the customer. An example of a special housing needs household which might require enhanced housing management support is a frail elderly person, who may require the services of a sheltered housing officer, or 'warden', either on-site or on a peripatetic – that is, a visiting – basis. Typically, a warden might liaise with medical personnel in case of accident, or organise social events in a sheltered housing scheme, so-called because the scheme itself offers a degree of security and support not found in independent accommodation. Other special needs customers, living in a housing scheme or receiving support in the community in their own self-contained homes, include households containing members with a disability which makes them less mobile than others, and who therefore need housing which may be suitable for wheelchair access, or ramps to enable access, exit and moving around the dwelling. It is worth dwelling on the sort of customers who might need enhanced housing management services. Consider the list in Table 1.2.

Since the mid-1980s, the trend has been to try to move away from an institutional approach to the containment of special needs groups, towards 'care in the community',

Table 1.2 Special requirements

Customer	Possible special needs/characteristics	Possible housing response
Frail elderly	Failing health: poor mobility; hearing and sight disability; confusion; loss of (or diminished) social networks	Place in a sheltered housing scheme with on-site housing management and possibly care support: visiting warden
People with learning difficulties	Inability in varying degrees to cope with looking after themselves in domestic, economic and social terms	Place in a supported scheme with on-site counselling, warden assistance, etc., or similar range of services provided in customers' own home on a 'care in the community' basis
Substance and alcohol abusers	Inability to find housing due to socio-economic problems, and need for ancilliary and perhaps non-housing care support in a safe and secure environment	Place in a dedicated hostel with specialist counselling and sometimes medical staff to help rehabilitation, or similar support provided in the wider community

which means bringing services to the customer in their own home. This policy was developed for two reasons: first, to reduce the dependence of people on institutional care in (for example) mental hospitals and geriatric wards in the belief that such institutionalisation tends to work against the possibility of eventual integration into the wider community, where this is at all feasible; and second, because it was thought to be cheaper to provide visiting services albeit through a number of different agencies that running institutions to contain them. It is true that there were short-term financial gains of this policy to the national and local state, with the sell-off of institutions generating significant receipts, but the reality is that providing dispersed care packages is frequently more expensive than providing everything on-site, as much of the element of economies of scale is lost.

A distinction is frequently made between housing and other forms of need in the context of providing for special needs groups, especially when the matter of how to pay for the services is considered. Non-housing needs activities include medical assistance, such as the administration of medication or therapies; counselling related, for example, to the control of drug or alcohol use, or behavioural problems; and social-work-type activities such as assistance with child-rearing or socialisation skills. Housing needs activities include assisting the customer to understand their tenancy agreement, with paying the rent and getting repairs done. This becomes very important for those responsible for the management of sheltered housing schemes when trying to recharge costs to Social Services, and avoiding loading non-housing costs onto the Housing Revenue Account. Social Services are paid for through the Revenue Support Grant – assisted general fund of councils, which collects council tax from the whole of the eligible community to fund services which are needed regardless of tenure, and the Housing Revenue Account is supposed to be a landlord account receiving rent and related income from tenants as tenants rather than as persons needing a non-housing service such as education, and spending on housing owned or managed by the council to make sure that it is maintained properly, and that tenants abide by their terms and conditions.

The *Local Government and Housing Act 1989* tried to draw a line between housing and non-housing activities ostensibly to stop councils subsidising rents from the rates (now council tax), including payments from people outside council tenure, although in some cases it notoriously required councils to pay money from its landlord account into the general council-tax fund and effectively subsidise housing benefit and other services or payments which are enjoyed by a much wider class. This debate will be continued in the housing finance chapter; the overall point is that this distinction is made for financial as well as organisational reasons, and it is often difficult to see where one set of needs ends and another begins.

Acute housing need

Economic need has briefly been touched on, and the effects of inequities in income distribution in relation to the cost of market accommodation is illustrated by a number of related phenomena, which vary spatially and over time, at various different scales.

One of the most poignant indicators of economically generated need is homelessness, the tackling of which is an enduring feature of housing policy at local and national levels. It even has its own legislation, to help define what the state thinks of as homelessness and which sort of homeless households deserve what level of help. The growth in the use of temporary accommodation, especially in London, has imposed serious and rising costs on the nation, as well as being unsatisfactory for those who have to live in it. It is always a area of controversy, and often the subject of moral debate, especially when the subject of discussion is young mothers or refugees and asylum seekers.

Homelessness is not a new phenomenon: people have always lost homes through war, fire, natural disaster, family dispute, financial difficulties and so forth, but societal attitudes towards homelessness have changed over the centuries. In Victorian times, the Poor Law dictated that homeless and destitute people would only get any form of help if they approached the great and the good of the parish of their birth or long-term residence, and sometimes not even then, with families being split up or being left to subsist in slums or on the streets. Charities were established to help deserving destitute cases from Elizabethan times onwards. Even in the mid-twentieth century, homelessness was frequently regarded – at least by those who were adequately housed – as a social disease, or as a personal failing, to be treated or regarded accordingly. It was only from the mid-1970s, which heralded the *1977 (Housing Homeless Persons) Act* that official attitudes towards homeless people changed. This Act, campaigned for by Shelter, a national charity aiming to fight the corner of people without homes or who are badly housed, imposed a duty on councils to house those who were homeless, had a priority need (for example, were pregnant, had children, had a disability or illness which would be made worse by being homeless), were not intentionally so (not having done something to cause their homelessness), and who had a local connection (if any) with that council through residence, job, family or for some other reason. Moral censure was imposed by granting those who satisfied all of these criteria except the non-intentionality clause temporary housing only to give them a breathing space to sort their affairs out.

There has been little change in state attitudes towards homeless households since then, apart from a widening of the priority categories to include more single-person households and lately certain categories of asylum seekers, and an attempt to reduce the use of bed and breakfast hotels as temporary accommodation by encouraging councils to enter into leases on private property directly or through RSLs and to move homeless families to lower demand areas.

The figures do not make pleasant reading: Table 1.3 and Figure 1.2 show how homelessness has risen year-on-year since 1997. The use of temporary accommodation, at enormous expense to everyone, has also risen, as has the backlog, as councils have found it ever more difficult to find appropriately sized permanent housing either directly or through RSLs to discharge their duties.

The problem varies spatially: predictably, high-demand areas such as the South East and London, characterised by

Table 1.3 Homeless households in priority need accepted by Councils, England, to 2003 (Source: Data from Housing Statistics 2003, ODPM (2003) table 621, p. 115)

Year	Homelessness acceptances (priority need)
1992/93	136230
1993/94	125360
1994/95	116850
1995/96	116550
1996/97	110810
1997/98	102430
1998/99	104260
1999/00	105580
2000/01	114670
2001/02	117840
2002/03	129320

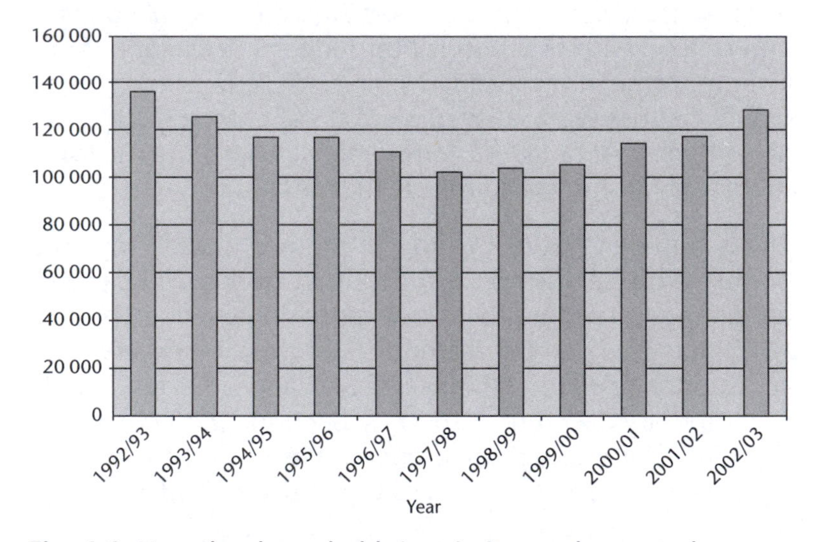

Fig. 1.2 Homeless households in priority need accepted by Councils, England, to 2002/03.

relatively high prices for owner-occupied and privately rented housing in relation to incomes, have seen higher and rising levels of homelessness compared to lower-demand areas, showing that it is not fecklessness but simple economics which is the cause of most homelessness – being in the wrong place at the wrong time.

The cost of homelessness is both a national scandal and a drain on public resources. A report published by Shelter, the national homelessness and housing aid charity, in 2004, indicated that there were 100 000 families living in temporary accommodation in June 2004, costing an estimated £500 m a year. The human costs are less easy to calculate. The same report indicated that two-thirds of the families lacking permanent housing stated that their children had problems at school, and claimed that thousands of children in temporary accommodation were absent from school for more than ten weeks a year. The knock-on effects for society, as a result of childhood deprivation, are incalculable (Shelter, 2004a).

It is, of course, true that there will always be non-explicitly supply–demand and economic causes of homelessness such as family dispute, fire, natural disaster, domestic violence and racial harrassment driving households from their homes, but supply problems are a major contribution to homelessness. The link between housing demand–supply imbalance and homelessness is obvious and compelling.

It is a mark of the seriousness and cost of homelessness to the state that the Office of the Deputy Prime Minister (ODPM) established a specific division to focus on the issue, the Homelessness Directorate, in 2002, and it has published advice to local authorities on how to prevent homelessness, following the refinement of legislation in widening the definition of vulnerability, as mentioned previously (ODPM Homelessness Directorate, 2003).

Overcrowding is also evidence of unmet housing need. It is a subject which once more came into prominence after a lapse of at least fifty years in 2003, when the results of the *English House Conditions Survey* were published, and linked to the 2001 Census figures. Overcrowding occurs most commonly where a household is unable to adapt to its changed size by buying or renting a more appropriately sized property in any sector.

The national overcrowding trend has risen disturbingly over the last few years, and it has been more accentuated in high housing demand regions such as London and the South East. London Housing, a local government agency which

Table 1.4 Overcrowded households in England by tenure, 2002/03 (Source: Housing Statistics 2003, from Table 8.6)

	Number of overcrowded houses in tenure ('000)	Total households in tenure ('000)	Percentage overcrowded
Owned outright	59	5 854	1
Buying with a mortgage	146	8 398	2
All owners	205	14 252	1
Rented from council	152	2 719	6
RSL	54	1 247	4
All social rented sector	206	3 966	5
Rented privately:			
unfurnished	44	1 473	3
furnished	42	563	7
all rented privately	86	2 036	4
All tenures	497	20 254	2

is part of the Association of London Government, which represents the interests of the 33 London boroughs to central government, published a briefing in 2004, which drew some depressing conclusions on the basis of figures in the 2001 Census. Amongst these was that overcrowding in London has gown significantly in London since the 1991 Census, almost doubling in the council sector, with severe overcrowding rising by nearly 50%. It pointed out that this growth reversed a long previous decline. Collateral facts included the shocking statistic that in 2001, nearly half a million children live in overcrowded conditions – around 15% of the capital's population. Overcrowding adds to existing disadvantage and deprivation in the council sector, and was shown to be thee times more common amongst black and minority ethnic (BME) households than white British households. This is a legacy of inadequate house building programmes, in terms of both number and space standards, since the late 1970s. The study also nailed the myth sometimes heard that moving tenants out of under-occupied housing would largely redress the problem of overcrowding in the social sector, showing that if all council tenants who significantly under-occupied their homes moved out, it would address London's overcrowding only to a very small degree. The report showed that the number of social housing households under-occupying by two rooms or more is less than half the number overcrowded by one room or more. A like-for-like comparison was not given, on the very reasonable assumption that many households would like to keep a spare room for guests, or for study, as is commonly the case in the owner-occupied sector (London Housing, 2004). This exemplification of overcrowding in a sector which was designed largely to overcome this phenomenon is surely an indictment of recent housing supply policies, as well as being partially a function of lack of affordability or accessibility to other housing sectors by households unable to exercise effective demand through income, social or cultural factors.

Overcrowding means that a household does not have enough space in which to live in a way which society expects: for example, there are too many people sleeping in a bedroom, or using the same domestic facilities and common rooms such as cookers or bathrooms. Overcrowding can cause mental and physical health problems, and has been shown to impede educational and social development amongst young people, as indicated in a recent report by the homelessness and housing advice charity, Shelter (Shelter, 2004b).

Therefore, indices of overcrowding should be very useful in determining where to build more affordable dwellings, and there is evidence that central government and the quasi-non-governmental organisation (QUANGO), the Housing Corporation, are taking more notice of these dreadful statistics when they allocate permission to incur capital expenditure or divide up the Approved Development Programme funds accordingly, and not before time.

The issue of key workers is of relatively recent interest to the state, but is not a new issue. The fundamental problem is that those wishing or needing to move for employment reasons often find this difficult or impossible due to an unavailability of suitable housing within their financial reach, and this has been the case for a long time. The issue came to prominence once more at the start of the new millennium, due to the identification of a shortage of essential public sector workers (now known as 'key workers') in London and the South East and the effects on the viability and sustainability of social infrastructure provision, such as health care and education. Various policies have been enacted, including the Starter Homes Initiative involving housing associations in acquiring properties for low-cost sale to certain categories of key worker, and the more widely scoped Key Worker Initiative, again enabled through RSLs and stressing low cost home ownership as the main way of providing affordable housing for certain public sector employees already living in or intending to take up work in areas where employment demand exceeded supply.

The discussion above illustrates the point that housing need and demand is often related to factors other than purely the availability or otherwise of housing, but that housing shortage can nonetheless be the cause of problems which go beyond just housing supply–demand imbalances. The central lesson here is that housing issues are rarely just housing issues, and should be seen in a holistic way to appreciate their true significance.

The management of housing need

Social housing is a response to quantitative and qualitative affordable housing shortage, and always has been, since Victorian times, when the first social housing was provided to alleviate health problems and in particular the spread of disease from insanitary dwellings to the better-off by providing an alternative to slum dwelling. In the marketplace, the allocation of scarce commodities is dealt with through the pricing mechanism: those who can afford to pay for the good get it, and those who cannot either have to adjust their bid or try to find a cheaper substitute, or otherwise defer or cancel their interest.

In contrast, the allocation of public sector resources, provided partially or wholly through the public purse, is done according to the need of households as perceived by the state. This is true for the range of public goods, payments and services such as health provision through the National Health Service, state pensions, social security, education and public transport personal subsidies such as concessionary fares for pensioners. The common feature of all of these forms of provision is that eligibility depends on individuals or households having to satisfy a range of criteria, either on the basis of having contributed towards a fund without a further means test, as is the case with National Insurance – related benefits such as the state pension, where eligibility depends upon NI contribution level and age, or through having a recognised need which cannot be met from the subject's own resources, which may not be contribution-based, such as 'free' school meals for the children of those claiming social security.

State benefits which are contribution-dependent but available to all who have paid in, regardless of current income or other resources, are known as universal – non-means-tested – benefits, and those which are not contribution-dependent but depend on assessed need only or mainly are sometimes known as selective – means-tested – benefits. Social housing is in the latter camp. From the early days, only those who satisfy certain housing needs criteria have been able to obtain social housing, although a feature of this benefit which is not generally true of state financial assistance is that beneficiaries are able to retain the gain even if they become able to satisfy their needs in the marketplace, as tenancies cannot be terminated on the basis of having breached a given affordable income level.

Social housing is rationed as a state-enabled commodity designed to meet housing shortages. Legislation prescribes the form of this rationing, principally the *1996 Housing Act*, which lays down the criteria which schemes devised to select tenants on the basis of need have to satisfy. RSLs have their own many and various means of selecting tenants, some based on the objects of their charitable status, others more directly on guidance issued from time to time from their regulator, the Housing Corporation. The severity of the gatekeeping depends upon the relative shortage of social housing of a given type in a geographical area: in some areas of the UK, it is hard even to give tenancies away even after extensive nationwide advertising.

Most social housing organisations have adopted a 'points system' or 'banding system' method of needs assessment to inform selection at some time. Crudely, points systems involve numerically scoring each applicant household for a given type of property against what are regarded as a set of key housing needs, informed by a combination of national and locally decided policy, and granting the tenancy to the household with the most points. The chances of selection are based on the relative position of households in the points league table, and change with every new application for a property of that category, usually defined by number of bedrooms or equivalent. Different selection criteria and weightings may apply when allocating properties with specialised features for clients with particular housing requirements, such as housing with care facilities. Household characteristics commonly assessed in this rationing process include:

- Medical aspects – is the present accommodation (if any) unsuitable because it makes a medical condition worse? (For example, the medical condition of someone with a heart condition living in a flat which can only be reached by steep steps may be worsened by having to climb the stairs on a regular basis.)
- Environmental health factors – does the condition of the existing accommodation fail certain public health standards? (For example, it may lack a kitchen or bathroom for the sole use of the household, or it may be damp, structurally unsound, or harbour vermin or other pests.)
- Overcrowding factors – is the accommodation statutorily overcrowded, so that the household does not have sufficient rooms to live a normal life; do children of different sexes and of a given age have to share a bedroom, are the rooms too small for their purpose in relation to the household's needs, etc.
- Social factors – is the household suffering harassment in the present accommodation which could be relieved by relocation; are essential facilities (e.g. schools, hospitals) too far away to be reached in a reasonable time; does the household need specialist support which can best be delivered in a scheme context, etc.

In the 1980s, quite a significant computer-based industry grew up which provided points-schemes advice to councils as supply shortages grew and the need for rationing increased. Before that time, largely due to higher levels of social housing production than at present, rationing of this complexity was less common, and in some areas, council housing was as much a tenure choice as any other form, including owner-occupation, and did not have any stigma of neediness. At that time, selection was more likely to be on the basis of time on a waiting list, and acceptable references from landlords on rent payment and abidance with tenancy terms than on the basis of need as such, although medical or social circumstances might have helped decide between people who have been on the list for a similar time.

Another approach is 'banding', where households judged to have similar levels of housing need due to a variety of factors (such as overcrowding, medical problems, mobility problems and insecurity of tenure) are placed in priority needs bands often designated A, B, C and so forth. Housing takes place according to the priority of the band and position, often related to date of registration. It would be particularly advantageous to be the applicant of longest standing in Band A – if a suitable vacancy comes up.

More recently, since the mid-1990s, there have been developments in the rationing of council and other social housing away from the standard points-system approach, to include an element of applicant choice, although need is rarely excluded from consideration, except in comparatively low-demand areas. One such development is that of choice-based lettings (CBL), loosely based on a model developed in the Dutch city of Delft in the late 1990s, where applicants with similar needs levels are often given a choice of properties advertised in local papers and on the provider's website, and encouraged to bid for them, with the household with the most points (as a from of cash) winning, replicating some features of competing for housing in the marketplace. Many councils have got together with RSLs in their area – sometimes on a regional or sub-regional basis – and operate a shared choice-based lettings system ranging across the entire social housing stock. This frequently involves providers adopting a common allocations policy, where the criteria for selection based on housing needs are more or less harmonised. Such schemes have met with some success, and customers appear to welcome this trend. However, in these cases there is no 'free for all' – the ultimate determinant of who gets the property is assessed housing need.

The ODPM has published a readable summary of CBL schemes, and applicant perceptions of them, in an attempt to get councils and housing associations to consider CBL seriously by 2005, and to adopt it by 2010 (ODPM, 2004a and b).

Since the mid-1980s, due largely to externally imposed financial restriction, local authority house building in the UK has been of negligible proportions. Government policy, directed towards maximising the input of private finance into the social housing development sector, expressed in the *1988 Housing Act* grant regime, has meant that the major producers of new social housing for rent and for sale are housing associations. This, combined with the effect of transferring homes mainly to custom-made housing associations under Large Scale Voluntary Transfer, since 1989, has increased the importance of nominations to local authorities in attempting to satisfy housing need. It is especially vital in areas of high demand and low turnover in the council rented sector, such as London and the South East. Most councils have negotiated nomination rights to housing associations operating in their area.

In this context, 'nomination rights' means rights to house a social housing applicant in a housing association property, be it a household which is already in council housing seeking a transfer to another more suitable property, or a household awaiting social housing and living in another form of tenure, or a household accepted as homeless. This can be contrasted with 'referrals', where a council informs a housing association that an applicant would prefer to live in a housing association property, but with no rehousing rights attached. The notion of nominations has been weakened, but not completely eroded, by the growing importance of common allocations policies and common waiting lists; and choice-based lettings schemes have reduced the need for nominations arrangements even further, as applicants

are effectively able to choose suitable properties within the entire social housing pool in which the system operates.

The level of nominations rights varies between councils and type of scheme. In some cases, where the council has supplied land at less than market value, where it has gifted the land, or where the housing association has received grant assistance to build the homes from the council under the now-abolished Local Authority Social Housing Grant (LASHG) system, or via other routes, it has been possible to negotiate 75–100% nomination rights on first lets to schemes. Very high nomination rights have also been secured where the land and/or property has been secured through *Section 106* arrangements (so-called 'planning gain') through a profit-sharing deal with a developer in return for the grant of planning permission on a piece of land not classed for housing use. The same is true for schemes built specifically to house homeless households, at least on first lets: and some councils have, in the past, negotiated very high nomination rights in perpetuity.

In other cases, councils have been able to negotiate lower levels of nomination rights, and it is very rare to find schemes where none exist. There are arguments for and against high levels of nomination rights. Arguments for include the fact that very many of those applying for social housing through local authority housing registers also appear on housing association lists, and therefore in many cases the household would have been accommodated in the same way had it applied only to the association, depending on the allocation criteria adopted by the association. Another would be that the association, having received a substantial grant to provide the housing, owes a public duty to house those most in need, and that councils, as statutory housing enablers who are publicly accountable bodies, whose allocations policies are prescribed largely by legislation (*1996 Housing Act*) are the correct conduits for such housing. Yet others include the view that, having received land or housing directly from a council, or indirectly through council negotiations, owe a reciprocal duty to that council to provide housing for its nominees. The pragmatic argument is simply that councils cannot cost-effectively develop their own housing, and can only satisfy housing need where there is low turnover and high demand by recourse to housing associations.

Arguments against high levels of nomination rights include the position that housing associations have their own constitutions with distinct aims and objectives, and should therefore be able to decide themselves who they house. Many associations pre-date council housing functions – for example, the charitable housing associations and trusts such as the Peabody and Guinness trusts, which were established in the nineteenth century, and have been housing people in need for far longer than councils according to their charitable aims. Very high nomination levels, unless all nominees' circumstances fall within the objects of the charity, would erode the independence of the associations in question, and could even mean a conflict with their constitutions. Another argument is that some associations deal with niche markets which are inadequately served by councils, such as young single homeless people, or households with specific needs who are not recognised as in priority by councils, and who otherwise find great difficulty in finding somewhere suitable to live.

The truth is that it should be possible for councils to nominate sensitively, taking account of the housing associations' many and varied aims and objectives, working in partnership; and it should be possible for housing associations to accept high levels of nominations where there is no good reason to differentiate between own-list and nominee candidates in terms of housing need. There will, however, always be dispute between councils and associations, unless some day a common allocations policy is imposed on both sets of bodies by legislation. Some associations have accused councils of dumping 'problem cases' on them, giving rise to higher than desired levels of arrears and anti-social behaviour, and some councils have accused housing associations of being far too restrictive on who they will house, and of 'cherry-picking' the most desirable cases, leaving councils with a rump of the most difficult households. Associations have been accused of simply not playing the game – of not abiding by established agreements – and it has in many cases proved difficult to monitor nominations rights, especially on relets. The National Housing Federation published an influential review of nominations policies, called *Changing Places*, in 2003 (National Housing Federation, 2003) which gives a clear account of the problems experienced in this area, and outlines some good practice approaches.

More recently, the emergence of regional-only Housing Corporation grant funding, replacing the allocation of SHG by council area, and the abolition of Local Authority Social Housing Grant (LASHG), has led to the establishment of protocols aimed at regularising nomination arrangements to new-build properties and subsequent relets between groups of councils and associations on a sub-regional or even a regional basis. Such schemes have in many cases been hard-fought, but there is evidence that this approach is working.

It may be that in the future, regional working will make smaller-scale nomination arrangements obsolete, which would be excellent for those seeking social housing, for whom suitable, affordable housing often ranks above precise location, especially in the context of continuous urban areas like Greater London, Merseyside and the West Midlands conurbation.

It could also be argued that it would help deliver the objectives of the *Sustainable Communities Plan* in ensuring take-up of homes in development areas such as the Thames Gateway, which might be shunned as a location by some social housing applicants, especially if there were to be some compulsion in restricting the number of offers which could be turned down before the applicant is removed from the register, although this is bound to be controversial, and may smack of social and spatial engineering.

Supply issues

Housing is supplied by private developers, councils and housing associations, for rent and for sale. As has already been shown, the heyday of direct council supply of new-build social housing has passed, and housing associations are the key provider of new social housing for sale and for

Table 1.5 New dwellings started, by year, United Kingdom, 1992–2002 (Sources: P2m returns from local authorities, returns from National Housebuilding Council (NHBC),National Assembly for Wales, Scottish Executive, Department for Social Development (NI))

Year	Private	RSLs	Councils	Total
1992/93	129 567	37 826	3 246	170 639
1993/94	150 707	41 472	3 058	195 237
1994/95	163 226	39 627	2 589	205 442
1995/96	140 467	32 892	1 657	175 016
1996/97	162 560	30 052	1 799	194 411
1997/98	171 634	25 414	1 011	198 059
1998/99	161 390	23 757	362	185 509
1999/00	169 305	22 478	444	192 227
2000/01	164 714	20 228	468	185 410
2001/02	177 337	17 499	193	195 029
2002/03	178 881	16 529	227	195 633

rent, although they almost always engage private developers to do the actual building. Table 1.5 above shows the new homes starts over the period 1992–2002, and it is evident that the majority of homes over that period have been supplied for owner-occupation through private developers, for direct sale.

With over 70% of households in owner-occupied housing, either owning outright or subject to a mortgage, this has been described by many as a natural form of tenure. It is true that survey after survey reveals that home ownership is what most households aspire to, although the degree to which this is the case is life-cycle dependent. For example, younger people in the earlier stages of household or family formation tend to be less inclined to this form of tenure in the short to medium term than other groups. This can be explained by greater job mobility, not wishing to be tied down with a large mortgage, not needing long-term housing, expectations of increasing household size in the near future and the perceived irksomeness and expense of buying and re-selling compared to the relative ease of giving up a private tenancy and finding another one. Despite this, the growing proportion of owner-occupation over the

…letions under this scheme over a ten-year …92/3 to 2002/03.

…he scheme has been to reduce council stocks, … housing associations, as the rights enjoyed by …nants are retained even where the ownership of …ties change through Large Scale Voluntary Transfer …nd many association tenants now have the analo-…ght to Acquire. Unfortunately, councils have been … to replace properties sold, due to inability to re-invest … of capital receipts from sales in building new homes, … to statutory restrictions which have from time to time …en reduced and sometimes increased, and so there has …een an absolute loss of council properties, especially of the …more popular types – family homes, generally houses, with two or more bedrooms.

Since the mid-1980s, councils have had to become enablers rather than direct providers, especially in areas where relets have diminished. Enablers are organisations which have to use other bodies to supply goods or services. An example of an enabler from outside the housing sector is a travel agent which secures holidays from providers like Sovereign or Cosmos, but who doesn't own resorts, commission flights, or acquire rights over hotel rooms through direct block-bookings with proprietors. Councils have increasingly moved towards a strategic housing role, involving drawing up rehousing plans and liaising with provider agencies like housing associations to rehouse people looking to the social rented sector. The most extreme form of this is where they have divested themselves of stock through LSVT, and rely entirely on relations with housing associations to meet housing need.

Areas of oversupply

In some parts of the country, and in areas within regions, cities and towns, there is no shortage of council housing, and in some parts there is a surplus of supply of social housing over demand. The reasons for this are several: some areas have lost their economic base, and are characterised

decades shows that owner-occupation is an increasing trend, and that it is likely that commercial suppliers will tend to cater for this established and enlarging market, increasingly leaving the supply of other tenure products to others.

This analysis is perhaps too simplistic. A house is after all a house, whether developed with the intention of sale or for rent in the social or private sector. A house can change its sector by being rented out by former owner-occupiers, or being sold to a sitting tenant: witness the Right to Buy which has to date seen the sale of 693 794 former council homes to their occupants between the financial years 1992/3 and 2002/3 (*ODPM Statistics, 2003*).

However, properties are usually constructed with a market in mind, and, to a certain extent, suppliers have actually skewed the market, rather than the other way about. If a developer will only build houses for sale, and can control the form of tenure through ownership of the sales route, then a given scheme will only be available for owner-occupation, at least initially. A developer may resist building houses which will be offered for private or social rent alongside dwellings built for sale, for fear that this will reduce the attractiveness of the properties to people who regard owner-occupation as a natural or superior form of tenure, which would affect demand and therefore the sales receipt.

Some inroads into the attempt by developers to reserve their schemes for owner-occupation on first transaction have been made, mainly but not exclusively through the town and country planning system. Planning policy guidance issued from time to time by the government has required planning authorities – county councils or unitary authorities – to have policies requiring developers to make a percentage of homes available as 'affordable housing' on a site-by-site basis. *The 1990 Town and Country Planning Act* 'Planning Obligation' regime has been used to negotiate the transfer of up to 50% of land or housing in new-build schemes to housing associations and other third parties in return for permission to develop on land outside the housing use-class, although increasingly the 'gain' is taken as a cash sum or as an offsite benefit due largely to developer resistance.

The buy-to-let marketplace grew considerably in the late 1990s and early 2000s, especially in response to a healthy demand from usually young professionals unable to afford to buy in their chosen location, or unwilling to commit themselves to a medium-to-long-term financial or maintenance commitment implied by home ownership. The market was stimulated largely by comparatively low returns available on traditional investments such as stocks and shares, or cash investments in banks and other financial institutions, due to a low interest-rate regime, and relatively high returns to be got from private rental income and property resale in high-demand areas of the UK. Predictably, investors tend to move in and out of the market, depending on the relative rates of return on alternative investments, made easier by the Assured Shorthold Tenancy regime, which means that houses can be sold on for owner-occupation with guaranteed vacant possession, and so this source of housing supply is not guaranteed either in terms of quantity or duration.

How do councils gain access to the housing provided by private developers, given that this is a most significant form of housing supply? One route is via nominations, as already discussed, to property constructed by private developers and purchased by housing associations. The housing can be in the form of housing for rent or for low-cost home ownership, under shared ownership deals or similar sub-market arrangements. Typically, shared ownership involves the housing association buying a property off the shelf, or getting a property built under contract, and selling a leasehold interest worth a share of the full sale value (anything from 25% to 75%) to a nominee of a council or to a direct applicant, and charging rent as consideration for the unsold share. There are frequently 'staircasing' arrangements, where the occupier can buy successive shares in the property until they become full owner-occupiers, whether or not the purchase is backed by a mortgage, involving the disposal of the freehold or under-lease to the occupier at the current market rate.

Another way of satisfying desire for owner-occupation, but which involves diminishing the supply of social housing by reducing turnover, is the Right to Buy. Some councils sold to sitting tenants before the 1980s, but the initiative was given

statutory backing from 1…
It involves selling counc…
been public sector te…
from £16000 (20…
subject to a ceili…
the UK, to ar…
cumulative le…

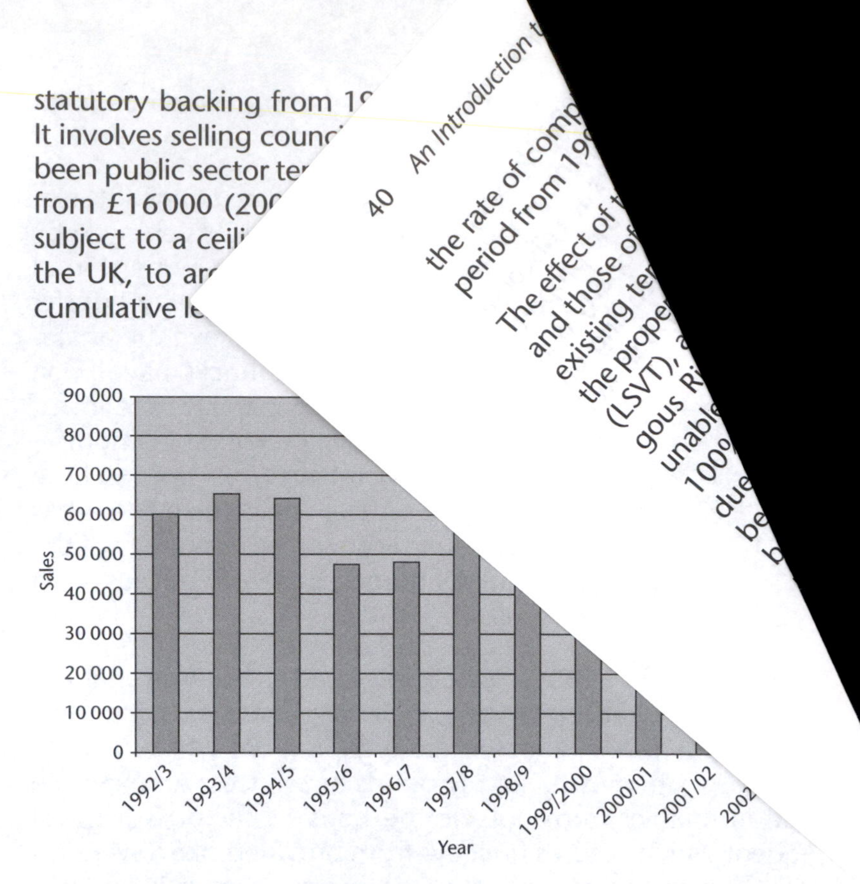

Fig. 1.3 Right to buy sales 1992/3–2002/03 UK.

Table 1.6 Right to buy sales 1992/3–2002/03 (Source: Local authority housing investment programme returns)

Year	Sales
1992/3	60 022
1993/4	65 275
1994/5	64 166
1995/6	47 304
1996/7	47 970
1997/8	58 746
1998/9	55 482
1999/2000	71 408
2000/01	69 049
2001/02	68 735
2002/3	85 637
Total	6 93 794

by high unemployment levels, prompting high level of out-migration, causing high vacancy rates in all sectors. Yet other areas contain house-types which no-one wants any more, and where households have moved to nearby properties which better cater for modern requirements, and where the social and economic infrastructure more nearly meets their demands. In some towns in the North East and Midlands, whole streets have been abandoned and rows of houses boarded up. In many cases, homes have been bulldozed, under the Market Renewal Policy, on the basis that it is cheaper to pull them down than keep them secure and maintain them in an empty state for eventual re-use. The problem is that houses are relatively immobile – you don't see many of them being taken down, packed up and transported along the motorway, although it is entirely possible to do so with some designs. It is easier and almost always more cost-effective to knock them down and build somewhere else.

High void rates cause problems for councils and other property owners. They still need managing and maintaining to a certain extent, while they generate no rent. Such properties represent a loss to the housing account. Areas with high void rates are often the subject of vandalism and antisocial behaviour, which make the properties even more difficult to re-let, and it is very difficult to reverse the downward spiral of decay. Councils have tried many methods to address this problem, ranging from demolition to trying to market the areas in conjunction with area improvement schemes, sometimes involving environmental improvements, conversion and repair, and economic renewal initiatives sometimes linked to area or regional economic re-development initiatives exemplified by Urban Development Corporation initiatives and Neighbourhood Renewal Areas which have an employment as well as a physical regeneration dimension. Such councils have enthusiastically embraced inter-regional mobility schemes such as the LAWN initiative started by the Association of London Government in 2002, and now part of the Housing and Employment Mobility Scheme (HEMS), which aims to encourage tenants and applicants in areas of high social housing demand to move to areas with surplus housing.

In low demand areas, nothing short of comprehensive economic regeneration will reverse the decline which market renewal schemes have tried to tackle. Unless or until this happens, the most appropriate solution is to knock the properties down and build new ones where they are needed. That is, if the space can be found.

Postscript to Chapter 1: Can regional planning solve the problems of supply and demand?

The *2004 Barker Review* identified a need for an extra 139000 new homes every year to 2015 in England, with 23000 extra social homes per year needed to that date – to meet current and projected levels of housing required, with most of these additional homes being needed in London and the South East, where the same report estimates the need for 32000 new homes, half of them affordable, every year, over the ten-year period. The 2004 Spending Review indicated that there should be an extra 75000 extra social homes for rent built between 2005 and the end of 2007, with production figures building up to around 10000 per year countrywide, in manageable stages, by 2007/08, boosting production by around 50% a year to just under 30000 (ODPM, 2004c).

The capacity for brownfield sites within cities has been examined, and found to be insufficient to meet this requirement. Planning restrictions on residential development in the Green Belt areas have been relaxed, and the *2002 Sustainable Communities Plan* identified growth areas earmarked for housing and related expansion mainly in the south of the country, including the Thames Gateway, Stansted, Ashford, Mid-Herts, and Milton Keynes.

Things must be desperate to contemplate building in a floodplain, in the teeth of government advice to councils in 2000 not to grant planning permission to developers who wish to build houses within the reach of river floodwaters for fear of non-insurability and the human and economic costs associated with evacuation and flood-damage re-instatement!

Planning, at the strategic and development control level, offers but a framework for development. It is up to private developers to take the initiative, and there are considerable obstacles to persuading them to play ball on the scale envisaged.

The infrastructural requirements of large development sites such as Thames Gateway – the cost of roads, flood prevention technology, and school and hospital provision, are considerable, and whilst some of the costs can be shared with developers, if willing, such initiatives require considerable public investment, as witnessed by the costs of installing the Limehouse Link in London's Docklands, at the time estimated as the most expensive road project in Europe. Then there is the saleability factor – it is one thing to attempt to build a way out of a housing shortage crisis, and quite another to market homes in areas with which people might be unfamiliar, or where they would prefer not to go: witness Milton Keynes, which has hitherto failed to reach its 250000 population target that should have been achieved in the 1980s, despite slick television marketing and one of the best shopping and leisure complexes in the western world.

Part of the answer may lie in successful marketing to change consumer attitudes. On the assumption that it is environmentally practicable to develop the Thames floodplain, and to develop perhaps 150000 additional dwellings in areas adjacent to marshland with apt names like Ebbsfleet and Greenhithe, and that a goodly proportion of these new dwellings will be affordable if not social housing, why not try to sell the area as an exciting waterworld experience? Homes on stilts as in Thailand to resist the rising tides in the wake of sea level rises caused by global warming; pontoon roads as in the Cays of Florida; a canal-boat quarter reminiscent of old Amsterdam, all served by water taxis: and the most superb marina imaginable.

Far more exciting, perhaps, than our usual vision of suburbia, and an idea whose time will come in estuarine Britain, perhaps.

Time will tell.

References

Barker Review Final Report – Recommendations (2004). HMSO.
The Communities Plan (2003). *Sustainable Communities; Building for the Future*. HMSO.
English House Conditions Survey (2001). HMSO.
Housing Statistics 2003. HMSO.
London Housing (2004). *Overcrowding in London*. ALG.
National Housing Federation (2003). *Changing Places*. NHF.
ODPM/Homeless Directorate (2003). *Reducing B&B use and tackling homelessness – What's Working: a Good Practice Handbook*. HMSO.
ODPM (2004a). *Piloting Choice-Based Lettings: An Evaluation*. HMSO.
ODPM (2004b). *Applicants' Perspectives on Choice-Based Lettings*. HMSO.
ODPM (July 2004c). *ODPM Statement on Spending Review*.
Shelter (2004a). Toying with their Future – the hidden costs of the housing crisis.
Shelter (2004b). Living in Limbo.

Bibliography and reading advice

Bevan, M. *et al.* (2001). *Social Housing in Rural Areas*. CIH/JRF.
Chartered Institute of Housing (2002). *Strategic Approaches to Homelessness. Good Practice Briefing no. 20*. CIH.
Fordham, R. *et al.* (1997). *Housing Need and the Need for Housing*. Ashgate.
Goss, S. and Blackaby, B. (1998). *Designing Local Housing Strategies: A Good Practice Guide*. CIH.
Rowe, S. with Cashman, C. and Zitron, J. (HACAS Chapman Hendy) (2001). *Partners in Strategy: RSLs, the Local Authority Strategic Role, and the shape of the sector*. CIH/JRF.
Wilcox, S. (2003). *UK Housing Review 2003/04*. CIH (updated annually).

Websites

It is advisable to try websites first rather than paper-based publications to find up-to-date facts and figures about supply and demand, although you should also read the weekly Chartered Institute of Housing-sponsored publication *Inside*

Housing, and its sister publication, *Housing*. The National Housing Federation's monthly magazine *Housing Today* is also a valuable source of information on how housing associations are trying to meet housing need.

The following are essential websites for statistical information on housing, including supply and demand.

- For census information: www.statistics.gov.uk/census2001
- For official housing statistics: www.statistics.gov.uk
- For official housing advice and factual information on all subjects relevant to local authorities: www.odm.gov.uk/housing
- For information about housing association supply and demand: www.housingcorp.gov.uk
- For the views of housing's professional body, the Chartered Institute of Housing, on supply and demand, and other relevant matters, access www.cih.org.uk on a regular basis.

2 Housing management

The management of social housing is a central housing function. There are two key aspects to it – first the management of people in the dwelling, and of the property itself. The first strand covers the management of tenancies, ranging over the collection of rental and other housing related income from tenants, enforcement of contractual and statutory tenancy conditions, ensuring that homes are properly looked after by those who live in them and agencies which own and have responsibility for them, and involving tenants in the running of their homes and the estates or areas they live in.

The second aspect covers the management of the property itself – in providing a maintenance service to ensure that the asset remains as a viable housing unit which can continue to be let to the existing occupants, or relet to another household if the tenancy is given up. Together, the two themes make up the central activity of local authority housing and housing association organisations. Other activities, such as strategy, policy and development, exist to plan for and to produce social housing, and are undoubtedly important, but the central aim of these functions is to produce homes for occupation under tenancies or on a leasehold basis, and the housing management function in its broadest sense can be carried out in respect of dwellings acquired from other developers on a purchase, compulsory or otherwise, long or short leasehold basis, or even on an agency basis for another owner and landlord. If it is not carried out effectively, the quality of the asset, wherever it comes from, and the social harmony of the neighbourhood, can suffer badly, and the work put in at the strategy and development stage can be all but undone.

It is true that poor design and infrastructural provision, and defective strategy decisions on neighbourhood allocation policies and tenure mix can jeopardise effective tenancy and property management, but most of these defects can be remedied by redesign and strategy adjustments. It is far harder to bring a neighbourhood into harmony again, or to rescue poor landlord and tenant relations and straightforward lack of trust as a result of lax or inappropriate management; and poor management can threaten the very viability of an organisation, especially if the result is a high level of voids and rent arrears which can threaten the survival of a social housing organisation dependent on a healthy income stream from rents and leasehold charges to run its services or to appeal to private lenders for further financial support.

As well as examining the dimensions of the housing management task whoever provides it, this chapter will also deal with third party management arrangements, in addition to the traditional direct method, where the owner is the landlord and provides the service. Until the 1980s it was uncommon to find some agency other than the owner of the property managing the tenancies, although there is a longer tradition of contracting out all (or aspects of) maintenance to another agency.

In the early twenty-first century, in addition to direct management, a variety of other models exists, ranging from management through a housing company set up under the provisions of housing legislation, management of council housing by housing associations, and in some cases, tenants managing their estates or street properties under contract to council or housing association, management of social housing by a private contractor, or a variety of methods mixing all of the above.

There have been significant changes in housing management styles over the last hundred years. In the nineteenth century, the methods of Octavia Hill were widely adhered to by philanthropic social landlords. Housing management was, with notable exceptions, practised on behalf of tenants, proceeding from a view about how people should conduct themselves in their homes, a view often at variance from the those of the occupants, and as much about social education and health as about collecting rent and ensuring that the property is kept in good condition.

Today, much of this has been turned on its head, with tenants having a far greater say in how their homes are managed and maintained, to the extent that many have taken over this function to varying degrees; and the notion of managing in the best interests of tenants without reference to their views and aspirations is all but dead. The reason for this change has much to do with changes in overall societal attitudes and the disappearance of the class system, or at least far greater social mobility, and the growth of cross-societal norms following largely from mass-media exposure and emancipatory educational practices, and arguably arising from greater sexual equality. The growth of a consumerist culture has also played its part in reshaping the ethos of housing management.

More specifically, the evolution of housing management must be seen in the context of policy and practice changes in the public sector as a whole. Fifty years ago, the idea of any recipient of public sector services being a customer with the right to specify, intervene in or modify the way in which public goods or services were provided to them would have been more or less unknown. Those in receipt of benefits, council housing, publicly funded medical treatment, education and social services were expected to comply with whatever regulations attached to the services in question; no satisfaction surveys or focus groups to find out how well recipients felt they had been treated, or what could have been done better then. Those seeking housing were 'applicants', benefit seekers were claimants, and occupants of hospital beds 'patients' – the common factor being that they were all passive receivers of services defined by others, politicians, bureaucrats and professionals – the great and the good – who knew better than those on the other end of the service what should be provided, and the level of service to be given. Unquestioning acceptance was the norm, with notable exceptions resulting from catastrophic or extreme events, such as the 1915 Glasgow Rent Strike, or the mass squatting of ex-military bases after the Second World War as a result of homelessness caused largely by war destruction. But the emergence of the consumer, complaint and litigation culture has blown all that away. The bureaucrats cannot get away with complacency about customer views and aspirations any more, even though there may be little alternative to what they provide for those who cannot buy their way out of social provision.

Hospitals are run by NHS Health Authority Boards who scrutinise customer satisfaction returns in judging the quality of medical care and management. Social Services departments are frequently in the media spotlight, and social policy is a political football. Customer panels are a universal feature of most council-provided services, and housing management performance indicators include customer satisfaction measures as key measures of effectiveness, affecting the level of funding available to such organisations.

The importance of customer views and involvement can be illustrated in the 1999 Tenant Participation Compact Initiative of the New Labour administration, which required every council to negotiate with its tenants on the extent to which they wished to be involved in running housing services and the standards they are entitled to expect, and the renewed interest of the Housing Corporation in the extent to which housing association customers are involved in the policies and practices of their landlord, and in particular, the extent to which tenants are represented on boards.

Another notable change in the nature of housing management has been its reaction, or rather adaptation, to demographic changes. The last few decades have seen an increase in the proportion of single or two-person households in the population, as well as its ageing. The birth-rate has been steadily declining since the late 1970s, and the 'traditional' nuclear family of mother, father and two kids has become much less common than it once was. Table 2.1 indicates

Table 2.1 Actual and estimated household number ('000) and size, England (Source: 2001 Census)

Year	All households	Average household size (persons per household)
1991	19 213	2.47
2001	20 750	2.34
2011	22 519	2.24
2021	24 000	2.15

the reducing size of households, even though the total number is growing.

These changes have varied spatially. London, the South East and the large conurbations have seen average ages decreasing, as well as a disproportionate increase in the number of small and often childless households, whereas rural areas show a more 'traditional' demographic profile, although populations here are older than the national norm.

In this context, the housing management challenge has been to adapt to the emergence of niche demographic communities. Many organisations specialise in providing housing for the active elderly, with an emphasis on a light warden presence with an assistance call facility in so-called 'sheltered' housing schemes, although many of these services can be and have been provided in the community without need for relocation. Yet others specialise in providing for young childless couples at the early stages of household formation seeking an affordable first home.

There have also been profound changes in the racial, ethnic, religious and cultural mix of communities looking to the state or its agencies or contractors for social housing for rent or for affordable sale. There is nothing new in diversity: but the scale of multiculturalism in the UK has indisputably grown since the 1960s, and is set to grow even more with the entry of central and eastern European nations into the European Union, with implications for levels of economically driven immigration and consequent need for an appropriate housing response. Housing organisations have emerged to address the needs of diverse customer groups, such as housing associations serving Asian, Chinese and Afro-Caribbean communities, in recognition of the overt and covert discrimination and disadvantage often faced by these groups, despite an ever-growing raft of equal opportunities legislation.

Women have been marginalised in western society in terms of political, social and economic power over much of the past two thousand years: but political emancipation following the First World War, and subsequent equal opportunities legislation in the workplace and in the context of matrimony and partnership, has increased their power and influence, so that the power balance between the sexes is more or less equalised. The consequences for housing management have been far-reaching, from the emergence of housing associations founded to provide decent housing for women on the basis of lower incomes though dysfunctions in the jobs market, to the almost total eradication of the view that housing management is essentially a female-centric 'caring' profession more suitable to the empathic tendencies of females, akin to the more traditional notions of nursing, espoused by Octavia Hill and her followers.

Lastly, a major influence has been the changes in macro-economic overviews of governments over the last few decades. 1950s and 1960s Keynsianism suggested that public sector expenditure was a desirable means of promoting economic growth and national wealth redistribution, and helped justify massive expenditure programmes including the provision of mass public social housing as well as a burgeoning cradle-to-grave welfare state. Housing management then was the management of a mass resource available to all who wanted or needed it, without being the preserve of the marginalised poor unable to get housing through any other route, without the strong social service provision ethos which goes hand in hand with much of the sector now.

Monetarism, based on the premise that inflation is caused largely by the volume of money in circulation, and must be controlled at all costs, was the ascendant star of the Thatcherite 1980s, and inspired significant public sector expenditure limitations on the basis that public spending was the one effective lever that government could control without jeopardising business interests by direct interest rate changes or credit controls. The run-down of many parts of the public sector, through Departmental Expenditure Limits and attempts to rein in Annually Managed Expenditure, the unruly part of current government expenditure including social security payments, varying with the fortunes of the economy in general and the boom–bust cycle in particular, have had their impact not only on the way in which social housing has been developed, and in the size of social housing programmes, but also on the way in which housing is managed. Macro-economic fashions have been largely responsible for a move towards out-sourcing council housing management functions and the formation of bespoke housing companies of management agents using private sector management methods and operating on reduced staff establishments, in a search for greater economy of delivery and reduced central government financial support for council housing management and maintenance functions. More will be said about these imperatives in Chapter 3.

Housing management, then, has been affected by a variety of influences, and there is little sign that it, or the forces which drive its evolution, are about to cease changing. These are relatively exciting times in which to deliver management services to residents of social housing.

General needs housing management

There has been a wholesale move away from paternalistic modes of housing management towards taking a lead on the form and content of the activity from customers themselves, and from other non-housing operations. Many working in the field in the 1960s or even 1970s would perhaps find it hard to recognise or identify with the style of housing management adopted by so many social organisations in the early twenty-first century.

The changing nature of the legal forms of tenancy provides ample evidence of the move towards a more customer-based culture in this area, as does the increasing involvement of tenants themselves in housing management, helping to drive a changing agenda, although it is essential to step back from the mass of innovatory practice to found an evaluation of changes on the basics of housing management which have already been identified.

Forms of social housing tenancy

Before the *1980 Housing Act* introduced the Secure Tenancy for local authority and housing association tenants, residents of council dwellings relied on contractual law and the complexity of landlord–tenant legislation scattered across the statute book for their rights. Whereas the private sector already had a relatively clear, if somewhat archaic, framework for landlord–tenant regulation in the controlled tenancy and (as a result of the *1977 Rent Act*) the regulated tenancy regime, there was nothing comparable in the social rented sector. Tenancy agreements varied considerably between organisations, and in many cases the occupants of council dwellings in particular had few rights beyond those of licensees; notice to quit was frequently sufficient to secure eviction, and there were little or no rights of appeal against a landlord's decision on terms and conditions, who could live in the property, or rent levels.

The *1980 Housing Act* arose through considerable pressure and debate amongst the housing and allied fraternity, both within and outside officialdom. Shelter, the national campaign for the homeless and rootless, was a key player in bringing the issue of public and voluntary sector tenants' rights to the table, and in securing a full set of rights and obligations which came in in 1981 as a result of the new Act. Key aspects of the new Act included:

- security of tenure rights, with clear grounds for possession, most of which were non-mandatory, based loosely on the regulated tenancy regime of the *1977 Rent Act*;
- a clear set of landlord and tenant rights and obligations, defining the scope and powers of housing management for the first time;
- right to a 'fair rent' for housing association tenants, determined independently from the organisation, again along the lines of the *1977 Rent Act* provisions;

- the right to buy for council and some other public sector tenants, at discount, based on length of tenancy, building on voluntary sales schemes established by some councils, notably Wandsworth Council in South West London.

The Act has been modified since, and consolidated with the *1977 Housing (Homeless Persons) Act* and other key housing legislation under the umbrella of the *1985 Housing Act* (now under the *1996 Housing Act*), which introduced Section 27 – the duty of Councils to consult with their tenants when proposing a major change in housing management. The latter is a key requirement which has played a major part in ensuring that tenants have been consulted on major developments since, such as transfer of management responsibility under 'externalisation' contracts to housing associations, specially crafted housing companies such as arm's length management organisations and to co-operatives formed of tenants themselves.

The council housing management task has been further defined by the *1989 Local Government and Housing Act*, which requires councils to publish clear performance indicators to show their customers, including tenants, leaseholders, the general public and the government, how they are doing against in most cases self-set targets on rent arrears collection, repairs turn-round time, and lettings to different sections of the local community, amongst other facets of the task. This requirement has been further refined over time through the best value regime and developments following the *1996 Housing Act*.

The diversification of council housing management, notably by empowering tenants to run their own show to a degree, was given impetus by the *Right to Manage* clauses of the *1993 Leasehold Reform, Housing and Urban Development Act*. This enables council tenants who wish to do so to manage their own homes, as long as 25 or more tenancies are involved. Effectively it involves creating a management organisation (a tenant management organisation, abbreviated to 'TMO'), directed by a board of secure tenants and often employing professional staff, to provide housing management services to the properties and tenants within its remit and also in some cases services to the community, such as litter and abandoned vehicle removal, under contract to the council which pays the organisation an allowance reflecting its own budgeted expenditure. The process of negotiating the contract with the council is supported by 'Section 16' funding paid under the *1986 Housing and Planning Act*, through the ODPM, to which the council also contributes.

The development of TMOs, which followed on from the highly successful programme of fostering local authority tenant management co-operatives during the 1970s and 1980s under a variety of model agreements, represents the single most successful housing involvement policy initiative in council housing yet seen.

Since the early 1990s, there have been relatively few significant legislational spurs to council housing management practices, although there has been plenty of government advice in the form of guidance and circulars, and some orders amending previous law. The *1996 Housing Act* set out a framework for local authority allocations policies, but built largely upon what many councils had already incorporated into their priority needs 'points systems' in any case. It also made some minor changes to secure tenancy rights.

Perhaps its most dramatic impact came in the form of introductory tenancies, coming in the wake of mounting concern over anti-social behaviour, allowing councils to grant tenancies with fewer security rights for a period of up to a year, allowing them to terminate the tenancy relatively quickly without enormous burden of proof, if the tenant failed to live up to his or her obligations. This supplemented the 'non-secure' tenancy regime which emerged from the *1980 Housing Act*, allowing councils to grant tenancies of less security to households given temporary accommodation under homeless persons powers and duties, and for similar reasons, also allowing relatively rapid eviction. Introductory tenancies do not have to be granted but are not surprisingly popular with council landlords!

There is continued debate about the ideal formulation for the council housing tenancy, and much of it revolves around how best to manage properties. The mark of success would be the more effective engagement of the end customer in the process of evolution and evaluation.

Housing association tenancies have also changed over the past decades. The *1980 Housing Act* introduced secure tenancies to the sector, with fair rents, and superior conditions of tenure to previous models. Significant change came in 1989, as a result of the *1988 Housing Act*, which introduced assured tenancies for new housing association tenants, whilst those created before its introduction remained secure. The underlying reason for this change was the change in the development grant regime: housing association grant was fixed prior to development, with the remainder of the cost of development being financed through private loans. Rents now had to be set to cover management and maintenance costs, and also development loan repayments. This new reality meant the abandonment of pre-set fair rents in favour of levels required to ensure that the outgoings implied by borrowing and running costs were met, which were in many cases scheme-specific or pooled across post-1988 Act developments by the association, and could no longer be determined or regulated formulaically on the basis of affordability or in comparison with other similar lets, by the rent officer. Assured tenancies, where rents were no longer subject to such regulation, allowed associations to set rents implied by the commercial realities of the new financial regime, and enabled the Housing Corporation to set grant rates in accordance with national macroeconomic constraints without worrying about the restrictions of fair rents. The new rents were in many cases higher for similar pre-*1988 Housing Act* properties, which gave rise to some controversy, not least amongst voluntary sector tenants. In many cases, it proved difficult to manage arrears, and to explain the rent differential, where management services were identical between properties developed under the two differing regimes.

Assured tenancies were more onerous for tenants than the previous model, with mandatory possession grounds, including that for rent arrears. No attempt was made to harmonise statutory rights across the social tenures – no right to manage or right to buy for housing association tenants, and very limited rights to information or consultation.

Many associations tried to level the playing field somewhat by contractually enhancing the tenancy, and not using the mandatory possession ground, and during the 1990s attempts were made to equalise secure and assured rents, usually by increasing secure tenancy rents, made possible by the abandonment of rent officer regulation for secure tenancies. Some associations allowed tenants to buy at preferential rates, or gave grants to enable purchase in the open market.

In the management field, many associations responded to the right to manage in the council sector by encouraging groups of tenants to set up co-operatives or estate management committees to give them greater involvement, and some encouraged greater control through promoting tenant membership of housing association boards. The Housing Corporation introduced a more rigorous system of performance monitoring and tenant satisfaction measurement in the late 1990s, and gave guidance to associations on how to involve tenants from differing cultural groups in decision-making and management. The *1996 Housing Act* introduced a right to acquire for tenants of non-charitable associations similar to the right to buy without statutory discount arrangements, as an alternative to other low-cost home ownership initiatives developed partly to offer association tenants affordable home-ownership alternatives, to free up rented stock. But association tenants do not have the same range of rights as council tenants to be consulted on important changes to management, nor do they have the right to manage their own properties, and the assured tenancy regime is very much less empowering than the secure tenancy one.

Housing associations can grant probationary tenancies (like the council introductory tenancy) and have used this form of tenancy in the same way as councils, to try to ensure that new tenants are suitable additions to the communities in which they are housed.

Change is afoot, with the proposed combination of assured and secure tenancy forms for the social rented sector, and presumably the rights and obligations, into the 'Type One' tenancy, with most private sector tenancies being subsumed into the 'Type Two' arrangement. If performed successfully, this should help equalise the management involvement and other rights for all social housing tenants, and will be a mark of progress.

Managing the patch

Housing management takes place over areas with differing geographical characteristics, and applies to many different spatial arrangements and designs of homes. It also takes place against a background of cultural and ethnic diversity, and has to be sensitive to variations in age, sex and income level, as well as to the differing aspirations of tenants individually and collectively. It also has to respond to the challenge of having different tenure mixes in the same area, with leaseholders living next to freeholders and tenants, a challenge exacerbated by the Right to Buy. There is therefore no adequate 'one size fits all' approach to effective housing management.

The challenge of managing housing over a local authority area has been met by a variety of approaches. Some councils continue to adopt – or have returned to – the centralist model, where housing management officers operate from one building in the city and visit neighbourhoods or areas as needed, and others have neighbourhood or area offices in distinct localities based on village areas or wards, or arbitrarily defined patches, from where a local service is delivered. That service is delivered either on a comprehensive basis, with a full range of housing management and advice services, sometimes in conjunction with other departments such as social services, or on a partial basis, with some things (for example, housing aid and advice, lettings and allocations, Right to Buy processing or leasehold management) being retained in a central office. Much depends on the geographical characteristics of the district or borough: a rural or semi-rural district with perhaps one or two main towns might lend itself to a comprehensively or partially decentralised service, whereas a district covering one town or city, where Right to Buy has significantly reduced stock numbers, might be more economically or efficiently managed from the centre. Some authorities have abandoned decentralisation due to loss of stock, and cost constraints.

There is no right answer to whether it is best to centralise or decentralise: everything depends on local conditions and what tenants and councillors want to see. There is, however, some evidence that locally based housing management with tenant involvement can be more effective than more central forms in getting arrears down or reducing anti-social behaviour, evidenced partly through the enthusiastic take-up of tenant management organisations, the sustained popularity of the co-operative movement, and the resurgence of neighbourhood-based management initiatives on the part of housing associations.

There goes the neighbourhood

The challenges of managing a housing patch are legion. In the ideal world, a non-interventionist strategy would probably be best, with tenants left to get on with their own lives, paying their rent, requesting repairs on the basis of the tenancy agreement, managing the behaviour of their children and co-operating with social services and the police in helping to control the wilder excesses of younger people and more disruptive elements. In such a world, housing management would consist largely of letting properties, visiting to ensure that tenants have settled in and are aware of all the services on offer, advising them on the most convenient way of paying rent, giving advice on tenancy conditions, making sure that repairs have been carried out properly, and ensuring an orderly move-out and relet where necessary. This is quite a long list, but put like this, it sounds quite a pleasant occupation.

Unfortunately, the real world is not quite so accommodating. There is no 'typical' estate or management patch, just as there is no 'typical' community anywhere. Some estates, or more accurately, areas of estates, down to street or block-level, be they council or housing-association managed, or dominated by owner-occupiers, are run-down hell-holes which are shunned by people living outside their confines, and hated by those unfortunate enough to reside there. This is often despite social and physical regeneration and other well-meaning initiatives. There is surprisingly little correlation between built form and the intensity of management difficulty, although too often the stereotype is of the 'problem' inner- or edge-of-city high rise block or walkway maisonette warren, with more than its fair share of drug abuse, criminality and plainly anti-social behaviour.

One 'problem family' or youth gang can mark an area as undesirable, have severe knock-on effects, to the extent that the area may be shunned by applicants, and develop high void and protest rent-arrear levels. Area reputations like this are often ill-deserved, but very difficult to shake off. Reputations are often invisible to the casual observer: I have often visited 'cottage estates' where there is little sign of vandalism or dereliction, bar the occasional abandoned vehicle or fired garage in a block, and have been told that the area is a 'no go' area after nightfall due to high levels of youth crime, or that people not from the dominant ethnic culture (if any) fear to venture there alone at any time.

Social malaise such as high crime levels or generally anti-social behaviour, car racing, drug abuse and overt gang violence and intimidation can hit any area any time any where and is not the preserve of council or housing association estates, nor is it the preserve of inner-city areas or edge-of-town anonymous suburbs. However, where they afflict social housing areas, they become housing management issues which must be dealt with, although housing management-based initiatives are rarely in themselves sufficient, although necessary.

In recent years, there has been the development of the use of anti-social behaviour orders, following the powers contained in the *1996 Housing Act*, (as modified by the *2003 Anti-Social Behaviour Act*) several criminal justice acts, parliamentary orders and guidance. These are orders granted by or on behalf of the court which are binding on the offenders, who can be minors, which can prohibit certain actions (injunctions) or restrict the offender to an area outside the place where anti-social behaviour was committed on pain of a custodial sentence or fine (exclusion orders). Acceptable behaviour contracts have also been used by housing organisations and the like to try to persuade offenders to modify their activities, with the promise of more severe action if such contracts are breached.

To be successful, these initiatives require the co-operation of a range of professionals, often locally based, including police, social services, probation, youth worker services, and housing management staff, as well as that of the perpetrators and their parent or parents (if relevant). Specific housing sanctions exist, including eviction in case of conduct affecting the quiet enjoyment of neighbouring homes and damage to and misuse of housing property, but this can result merely in moving the problem elsewhere and needs to be backed up with more generic approaches.

It is also important to tackle the root causes of area degeneration, which goes beyond housing management. It may be that years of under-investment in the physical fabric of the dwellings and environment has led the area to become severely run down and that some local people, believing themselves to have been marginalised, have reacted by behaving badly or by letting children run riot. Part of the solution may lie in physical regeneration, coupled with 'planning for real' exercises with local people, helping to make something better of the local environment and giving local ownership to the remodelling. Again, the problems of an area may lie with high unemployment levels and consequent above-average levels of social security dependency, in which case an appropriate approach may involve locally-based employment and training initiatives.

Several housing associations, in an attempt literally to rebuild neighbourhoods, have set up or encouraged companies to facilitate the training of unemployed local people in building and allied trades, leading not only to physical regeneration and reconstruction, but also to individuals gaining marketable skills and entering the labour market, reducing the drain on taxation resources and helping to stimulate and sustain the local service economy. It may also be a factor in reducing poverty-related crime. Housing management has an important role to play here, in strategy, co-ordination and monitoring, but again, it is not, nor should it be, an issue to be dealt with solely by housing managers.

The control or management of anti-social behaviour is an area where joint working can bear fruit, and can foster co-operative strategies to the greater and more general good, as well as area management initiatives involving tenants in the driving seat, such as TMOs and area-based co-operatives.

Special needs housing management

There are some groups in society who are less able to secure accommodation in the marketplace than others, both because they may not have the financial resources to do so, and because the market is not tailored to providing for their needs. These are people who have been described as having 'special needs'.

The phrase is not fashionable in some quarters, partly because of the term 'special'. In a sense, everybody has special needs because they are unique individuals, but it is worth exploring what the term may mean in the context of particular groups and individuals which comprise them.

Needs in this context can be related to a characteristic or group of characteristics which leads to the requirement of particular types of housing and/or management support arising from their conditions. This housing is now known as 'supported housing' rather than 'special needs housing', as a result of changed perceptions already outlined.

The number of households with a disability or impairment which justifies the provision of adapted housing, either on a new-build or conversion/modification basis, is significant. In 2003, the *English House Conditions Survey* reported that in 2001, 270 000 households contained someone with a serious medical condition or disability that required special adaptations to be made the home, but their accommodation was not suitable for that person. Of these, 181 000 reported that modifications could be made that would make the accommodation suitable (ODPM, 2003a).

People with physical impairments or disabilities may have mobility and support needs which would not be the case for others, for example, standard house-types may not have corridors or doors wide enough to enable wheelchair users or those with walking aids such as frames to negotiate them and therefore to use the accommodation as people with fewer or no disabilities might. If housing with wider than standard door sets or with ramped or level access is not available, then they will be disadvantaged in relation to what others expect from a home. The extent of their impairment or disability might mean that they need some form of care support, which could be provided by the presence of a carer able to assist in the preparation of meals, cleaning, washing and other requirements, to enable them to live a life which is as close as possible to that enjoyed by those not in this situation.

'Impairment' is a term used to pick out a characteristic of an individual which limits the full use of a property, but not the degree of disability. For example, someone who has problems in using their arms due to muscular wasting or injury may be able to move around their home, or prepare meals, with difficulty, but reasonably effectively, given time or the presence of minimal aids and adaptation, such as ramped access or long-lever taps which do not require the same physical force to use as standard designs.

'Disability' is a term often used to describe the absence of radically reduced availability of a bodily or mental function, which means that adaptations of home design, and support, have to be present or given to a very significant degree. Both terms are ones of degree rather than absolutes, and the conditions they pick out may be permanent or temporary, and it is hard and perhaps not useful to try to define the boundaries of each ascription.

It can be argued that it is not the impairment or disability in itself which is the problem in use of accommodation, although it is clearly a contributory factor, but the limitations of the accommodation itself which disable and disadvantage, and the role of housing management in its broadest sense is to take away or at least mitigate that disadvantage. The correct focus of assistance may not, therefore, be the individual, but the housing or support which is provided.

On the basis of this brief discussion, is can be said that special needs arise where individuals or households cannot enjoy the same life chances, in this case specifically in relation to their use and enjoyment of their home, which others not in that position might if they choose to. The need, then,

Table 2.2 Client characteristics and housing response

	Design	Management
Physical characteristics		
Inability to fully use limbs to access, move around, or use accommodation or its facilities	Design to wheelchair or mobility standards Appropriate appliances	Carer assistance Specialist advice Effective inter-agency working
Mental characteristics		
Inability to reason sufficiently to carry out operations which lead to competent management of home and life activities	Simplification of appliances: care taken to ensure safe use of facilities	Carer assistance Specialist advice Effective inter-agency working

arises from a characteristic or characteristics of individuals. A brief (non-inclusive) classification of characteristics which might satisfy these criteria could be as Table 2.2.

Both physical and mental characteristics may be present to a varying degree, and the level of assistance might vary considerably.

In the past, policy responses were often geared towards placing people with some or all of these characteristics in an institution (institutionalisation) where all care and support could be provided on-site, rather than on an outreach basis, on the basis of administrative convenience or economy. That this form of provision existed (as it still does to a limited extent) gives the lie to a certain extent to the myth of the 'golden age' where families supported less able members without recourse to voluntary or state help, or where 'society' gathered its resources together to enable people to remain in their communities of choice. The model was that of the hospital, where the patient is brought to the institution to receive a package of care until well enough to go home, if ever.

Mental asylums of Victorian times, and geriatric wards, come to mind, as do care homes for the elderly and infirm. This was never a panacea, and everything depended on society's perceptions of the degree and type of need, and not everybody with such needs would have been accommodated in such institutions, or have expected to be so, and there was great deal of 'care in the community' even then, either through choice or design, but the approach was far more widespread than it is today.

Part of the approach stemmed from a view taken, largely amongst health professionals, politicians, and the 'great and the good', that the state or at least those with a controlling interest, in a position of relative power, knew best about the needs of those possibly more vulnerable than themselves, and from the view that it was cheaper to deal institutionally than on another basis. Some of its exponents would have considered it most irresponsible not to provide institutional care on the scale given, due to the effects of under-provision both on the 'patient' and on society at large, as evidenced by the number of charities founded by well-meaning people such as Shaftesbury and Barnardo, and by the church, to provide just such institutions, often as a product of 'Christian' duty or humanistic philosophies of responsibility towards weaker members of society. It is very hard to criticise the initiatives of a past age unless we can get into the mindset of those who founded or operated them, as the dynamics of that past culture have either gone or have been eroded to the extent that we may gaze upon their strategies with the same disbelief and lack of comprehension as when we visit the physical remains of the Mayan civilisation. Even reading the writings of the day give little illumination, as these are also artefacts only, and cannot recreate or reproduce the conditions which gave rise to them, although the novels of Dickens and the writings of the great social reformers of the day are illustrative.

Whatever its antecedents, the philosophy of institutional care continued until well into the twentieth century, under the aegis of the reformed Welfare State of Attlee and Bevan after the Second World War, right up to the early 1980s, when the whole idea of institutional provision on the scale seen was questioned, partly due to changing views on human dignity, and also due to macro-economic constraint, and a view that the state apparatus should be smaller.

'Care in the community' policies, entailing the closure of mental institutions and geriatric wards, and the scaling down of 'Part Three' social services care homes provision, came into vogue, and there was an increased emphasis on joint care planning in local authorities, where social services, health authorities and housing providers would work together to ensure that as many vulnerable people as possible, consistent with personal and community health and safety, would be able to remain in their own homes, or more accurately, in independent accommodation, with a package of care and support supplied to them there. That approach remains, although it has its critics, both amongst social policy commentators and sometimes the individuals and their carers themselves, but it is unlikely to be reversed. Indeed, the sophistication of policies and practices to try to guarantee high degrees of independence amongst those who would previously have been institutionalised has grown to the extent that such a reversal is probably as unnecessary as it is undesirable.

What, then, is the housing angle on all this? The first consideration is the design of housing, which will be dealt with more fully in Chapter 4. In the 1970s, enhanced wheelchair and mobility standards were brought into force for dwellings designed for household types with relevant characteristics, and have been refined over the years. Some RSLs now build homes to 'lifetime homes' standards to try to ensure that people do not have to move even if in the case of impairment or disability, a trend reinforced by Housing Corporation design standards. Much older stock has been adapted to such standards, either on a case-by-case or programme basis, if only because the possibility of supplying sufficient new accommodation to deal with the requirements implied by an ageing population is a remote one, given macro-economic constraints.

The second consideration is of housing management. With the reduction of 'Part Three' accommodation provided by social services for elderly or disabled people with multiple care needs, the social rented sector has had to take more responsibility for providing suitable homes with care and support where needed, which has meant a change in management approach. A whole industry has grown up to train and support 'wardens' or sheltered housing officers in providing housing management services to sheltered housing schemes, or on a peripatetic (outreach) basis, both in the local authority and housing association sectors.

The 'sheltered housing scheme' is typically a purpose-built block or group of dwellings designed with mobility needs in mind, and staffed by housing officers who, in addition to dealing with the standard housing management tasks such as rent collection (or more usually, advice on making direct rent payments through bank accounts), enforcement of tenancy conditions, and giving housing advice, may also act as an enabler for medical and social care provision via non-housing agencies.

Indeed, it is important at this point to make a distinction between housing and non-housing management or care issues. The rules governing the Housing Revenue Accounts of local authorities mean that in general, care or so-called welfare services cannot be financed from that account, which is used to employ sheltered housing officers, as they are not strictly 'tenancy' functions – to do with property management. If such services are provided through the HRA, there must be a recharge to another appropriate account. This follows legal judgements in the 1990s, famously the 1993 *Ealing Judgement*, where a specimen HRA tenant's complaint that HRA funds, derived partly from every tenant's rent, and government subsidy intended to support HRAs generally, were being used to pay for care services which would only benefit a certain class of tenants, and not to the benefit of all.

The central thrust of all this is that care services should be paid for from the General Fund of local authorities, which pays for services that anyone in the community could receive, generally or subject to their needs, such as refuse collection, education, roads maintenance, housing benefit, or social support services such as 'meals on wheels', reinforcing the status of the HRA as a ring-fenced (dedicated) 'landlord account'.

The other good reason for sheltered housing officers not providing medical or social services-type support is that it would be an extraordinary individual who could provide all to newspaper advertisements saying something to the effect of 'no blacks or coloureds'. Certainly before the Act, which made such forms of direct and more subtle indirect discrimination unlawful, this form of overt prejudice was relatively common, stemming partly through ignorance and consequent fear of difference, and partly through sheer unstudied hate and unpleasantness. Even people from black and minority ethnic (BME) communities with reasonable financial resources found it difficult to buy houses outside certain neighbourhoods, due to community resistance, the activities of racist parties such as the National Front, and the less overt but just as calculating policies of building societies who sought to 'red line' areas to exclude BME households from getting loans to buy in certain areas, leaving them to buy usually sub-standard or otherwise less desirable homes elsewhere.

Some local authorities and housing associations made things difficult, often unwittingly, by imposing local connection or length of residence qualifications on their housing registers, which newcomers to the area from overseas and those with no family connections could not hope to meet; and yet others tried to contain what they saw as a 'problem' by earmarking certain blocks or areas for BME housing, a form of apartheid only slightly less pernicious than that practised in the South Africa of Verwoerd and his successors from the 1960s to the early 1990s.

The drivers to such prejudice and discrimination were several, but related in the UK and other Western European states to the legacy of Empire and past conquest and alliances, and the exploitation of the peoples of colonial Africa and Asia and elsewhere. In the late 1950s, the then Home Secretary Enoch Powell urged citizens of Jamaica and other West Indian states which had the Queen as their monarch to move to Britain to take up jobs in essential public and related services such as nursing and bus driving, due to full employment conditions in London and the South East in particular, and the need to service the infrastructure to enable continued growth. Famously, a recruitment depot for London Transport was established in Kingston, Jamaica. Many took up the call and a promise of a new life, albeit expecting in many cases only to be resident in these damp,

of these services to the degree required, and it would be a very risky strategy, both in terms of health and safety of the employee and the person receiving services, to try to do so. Another thought is that to impose such a burden on one individual would be exploitative, given the general level of wages associated with sheltered housing posts. For all these reasons, sheltered housing officers, in the council and RSL sectors, act as facilitators and enablers, making sure that the right professionals in the health and care areas are available to give a targeted service to those living in schemes. Such officers may sometimes, however, be the first port of call in an emergency, such as attending to a fall, through an alarm or similar system, and are often trained to give basic first aid and assistance until emergency or other services arrive. Sheltered housing officers are best thought of as general-needs housing officers with some awareness of and training in social support and health care.

The services provided in a scheme context can also be given on an outreach or 'peripatetic' basis. Alarm systems installed in individuals' homes can be used to summon assistance; and call-centre approaches are common, which can direct officers to people's homes without the need for resident operatives. The effectiveness of remote services depends largely on the information and communications technology arrangements in place, and the willingness of individuals to use them, but, for all except those with profound disabilities and impairments, this seems to be the way things will continue to go.

The funding arrangements relating to providing housing and care support to individuals in schemes and in their own homes have changed significantly since the introduction of the 'Supporting People' regime in 2003, where payment for such support now goes largely to the landlord or care provider rather than to the individual through housing benefit or welfare benefits, and will be discussed at greater length in the next chapter. An excellent background source for information on the Supporting People regime is the ODPM's Supporting People review (ODPM, 2004b).

The need for supported housing, provided as a scheme or as a management initiative, will grow over the years, as the demographic profile of 'more economically developed countries' (MEDCs) continues to change, with the proportion of older people in populations continuing to grow. Although there is no one-to-one relationship between such an increase, and the numerical growth of disability or impairment, there is clearly a link, as the likelihood of these unfortunate characteristics does increase with ageing, with implications for increased supported housing provision and management, whether financed through state or private resources, depending on the fortunes of the economy.

Special consideration groups

Social housing organisers also provide for members of groups who find it difficult to access market housing for rent or for sale not through disability or impairment, or even necessarily for economic reasons, but who have been, and in many cases still are, subject to discrimination or disadvantage for a variety of reasons, but mainly centred on societal attitudes towards them.

Such groups include households who suffer economic and social discrimination and disadvantage through racial, ethnic or cultural status, groups marginalised because they have been or are disapproved of by those who they would otherwise look to for housing opportunity, and others whose lifestyle makes it hard for them to sustain tenancies or home ownership. The range is diverse, and there is no common factor other than the marginalisation already mentioned.

Concrete examples of such groups, in addition to those marginalised through ethnic, racial or cultural stereotyping, include (not an exclusive list – add your own):

Unfavourable medical or related condition:

- AIDS and HIV victims
- drug abusers and their dependants
- alcohol abusers

Possession of other characteristic(s) meeting disapproval and marginalisation:

- young lone-parent households
- young offenders (legal and non-legal behavioura
- time-served criminals
- street sleepers
- beggars

Victims of societal prejudice and indifference not itemis above:

- asylum seekers and refugees
- gypsies and travellers
- those discharged from the Armed Forces
- anyone who does not conform with prevalent societal view of 'normality'

Others who are hard to fit into the above categories, but who find it hard to secure mainstream market or social housing:

- students
- victims of domestic violence

This is a long list, and many people who fit into the above categories are in fact able to find housing either through market or social routes, but they are less likely to do so than others. A variety of housing providers has been formed to meet their requirements, most of which are housing associations.

Households who suffer economic and social discrimination and disadvantage through racial, ethnic or cultural status

Readers who were around before the *1976 Race Relations Act* may remember signs on properties for rent or qualifications

relatively cool islands for as long as they were needed, and many were allocated places in hostels and local authority housing, although others had to take their chances in relatively squalid private rented housing. Some fell foul of the notorious landlord Rachman in Notting Hill; others to the slum landlords of South London.

In all too many cases, the incomers were greeted with prejudice and anti-social behaviour on behalf of the 'host' population, and ruthless exploitation by landlords, which fuelled the Notting Hill Riots of 1958. When it was realised that many immigrants from the ex-colonies or New Commonwealth countries would not simply pack up and leave when they were no longer needed, they became the butt of even more vociferous prejudice, which the law did little or nothing to rectify. Enoch Powell made his famous 'Rivers of Blood' speech in 1963, in which he predicted wholesale social unrest and destabilisation if the 'tide' of immigration was not halted, or repatriation enforced, and the National Front blossomed and grew in its wake.

By the 1970s, racism was more or less institutionalised. It was only hard lobbying by the representatives of BME communities which managed to deliver the Race Relations Act, following hard on the heels of the *1975 Sex Discrimination Act* in 1976, and which at least formally outlawed direct and indirect discrimination against women. However, it took many years for things to get better in employment and housing terms for BME communities, and there are still profound barriers to equality of treatment in all respects.

Other challenges came as a result of the expulsion of the Kenyan and Ugandan Asians by dictatorial regimes and their seeking sanctuary in the 'motherland', only to find rejection and extreme prejudice awaiting them.

In response to this situation, several housing associations and co-operatives were established, including Ujima, Presentation, and ASRA, which prioritise BME applicants for good quality rented housing, staffed by personnel, often from BME communities themselves, who understood the stresses and difficulties caused by prejudice and discrimination, and who were able to give advice and assistance to BME households in order to help them obtain social justice through access to welfare benefits, educational and training opportunity, and in many cases, employment.

Today, the Housing Corporation has a well established BME Housing Policy, (Housing Corporation, 1998) and tries to ensure that part of the Approved Development Programme is made available to meet BME housing needs. It has a respected BME Code of Guidance which advises housing associations on ways of ensuring inclusivity, respect for diversity, and effective BME monitoring of lettings and transfers to ensure that no section of their applicant group is excluded for non-housing needs reasons. The Federation of Black Housing Organisations acts as a respected pressure group to maximise funding for BME associations, and there are very many training courses for housing personnel to ensure that they deliver policies geared towards equality of opportunity and non-racism.

Other immigration challenges, requiring a sensitive housing policy and management response, include:

- Vietnamese Boat People in the wake of the Vietnam conflict – early 1980s.
- Polish refugees in the pre-Solidarity era, and other refugees from Eastern Europe fleeing from dying state communist regimes prior to the revolutions of the late 1980s.
- Families fleeing the former Yugoslavia as the result of genocidal conflicts in Bosnia, Kosovo and elsewhere – mid- to late 1990s.
- Asylum seekers from Iraq and Afghanistan (late 1990s–early 2000s).
- Economic and cultural migrants from the European Union Accession countries, following EU enlargement, from April 2004.

Housing managers need to trained to be sensitive to cultural diversity and the variety of requirements of incoming people, and the need for this, and the promotion of housing bodies which can target their resources to making life easier for BME groups who make a tremendous contribution to the UK and EU economy, and to the cultural life of nations.

Empowerment and involvement of BME citizens in running and planning for housing was a theme addressed rather late in the day by the government in 2004, when a joint ODPM–Housing Corporation good practice guide on the subject was published, stressing the importance of engagement and suggesting how it should be done (ODPM, 2004a).

Unfavourable medical or related condition

Society regards some medical or related conditions in one moral light, and others in quite another, depending largely on the fashion of the times, and changing views on moral desert, and then not always consistently. Take drug dependency: cocaine and opium were the preferred recreational drug of Victorian upper-class society. Many royal figures were addicted to ambiguous substances, and Queen Victoria famously used cannabis to treat her arthritis and perhaps help her to come to terms with the death of her dear Albert. Shelley and Byron were significant substance users, and literary giants like Oscar Wilde and Aldous Huxley were not known for their abstemiousness. It has only been since the late 1960s that the abuse of heroin and other addictive and some non-addictive amphetamines has been stigmatised legally and morally, although it has to be said that the quantity of consumption of such substances, and their availability, has increased over the years. Despite the decriminalisation or softening of legal attitude towards some classes of drugs, users who find themselves dependent upon their support, or who need space, time and other resources to come off them, often find it very hard to secure suitable housing, either through economic problems, or because few want to let them into their home or property, for fear of criminality and anti-social behaviour, and to the stigma which drug abuse brings.

Some housing associations have grown up to cater specifically for drug abusers who are trying to come off their habit and re-integrate themselves into society, or where these people are significantly represented in their clientele – for example, associations catering from those who would otherwise be on the streets, such as St Mungos, and advice and assistance agencies such as Crisis and Centrepoint. They are staffed by personnel who have an understanding of the specific problems and challenges faced by such users, and who may even possess some medical qualification. Many of their developments are supported by Housing Corporation Social Housing Grant, and running costs by Supporting People monies from the state, as well as charitable donations.

Local authority housing managers have also had to adapt their practices to dealing with drug abusers and rehabilitees, as a result of the widening of the definition of vulnerability under the *2002 Homelessness Act*: an awareness of the specific and more general support services available is now essential, so that a package of enablement and direct provision can be offered and sustained.

Another group of sufferers who have been stigmatised in the marketplace, and who often need some form of supported housing provision, are victims of AIDS and HIV. The number of sufferers has grown dramatically since the early 1980s. Sufferers face the double whammy of loss of income and often mobility, together with the extreme prejudice often meted out to those who are believed to be or are in fact gay, due to the often incorrect assumption that AIDS and HIV sufferers are in general gay, that this is a lifestyle choice rather than a fact of sexuality, and that victims have in some way chosen their fate or made it more probable through their 'unconventional' activities. This prejudice may reveal itself in the termination of Assured Shorthold Tenancies where a landlord may otherwise have granted a new term, constructive dismissals, and the denial of accommodation by landlords through fear, ignorance, or downright hostility.

Many AIDS or HIV sufferers end up in local authority housing, either general-needs or supported, and many authorities make accommodation available above the bed number required (perhaps a two-bed flat where ordinarily a one-bed or bedsit unit would have been offered), under the direction of the *Homelessness Act 2002* or as a result of their points systems or banding schemes. This is to enable carers to be close, to offer the emotional and physical support for those who will often die as a result of increased vulnerability to infections and other forms of disease as a result of the weakening of their immune systems. Still others, including housing associations, employ specialist staff who can liaise with health authorities in providing a package of appropriate care.

Possession of other characteristic(s) meeting with social disapproval and marginalisation

The range of such groups is so vast that it would be impossible to do justice to the subject in a small library let alone a small book on housing policy and practice. They change with societal fashion and moral codes, and vary within and between countries and groups of states. The central distinguishing feature of this huge class is that at one time or another, the reason for their 'problem' has been identified as their own fault, and therefore not something which society at large should try to rectify, unless it is or could potentially be a threat to other relatively deserving or faultless members of society. The notion of the 'deserving' and 'undeserving' poor has been with us for a long time, and seems to be more or less a permanent feature of society, despite Judaeo-Christian teaching, or similarly derived high-level ethical codes.

Some individuals are probably best dealt with by exclusion: career criminals, rapists, murderers and violent thugs should probably be incarcerated, if there is no hope, on a case-by-case basis, of behavioural modification to neutralise the threat to society, or of ensuring that their energies are directed towards more positive activities. This is not the place for a full-blown discussion of the relative merits of retributive and rehabilitational justice, but when one moves away from the obvious groups, the rationale for social exclusion becomes rather more blurred.

Much of our ascription of 'less deserving' to certain groups seems to derive from a functionalist view of society espoused by *Durkheim* (1936) and others, who believed that the worth of individuals should be viewed in relation to their role in sustaining the organism of society, where the aims of society are greater than the aims of individuals who happen at any one time to comprise it. This is like the view that the interests of the company or organisation are superior to, and should dictate to a greater or lesser extent, the aims and actions of individuals making it up. This holistic view – literally that the whole is the greater than the sum of its parts – has its philosophical antecedents in Plato, Machiavelli, Kant, Hegel and Marx, to mention but a few great names, and feeds political philosophies and practices such as communism and socialism, and the view that there is a class of persons who know best, and who therefore have a right to define social and cultural values and control apparatus, runs throughout history, and is exemplified by Hitler's and Mussolini's fascists, as well as by the excesses of Stalin and Mao.

It is not, therefore, surprising that populist and frequently popular views exist which have the effect of excluding certain groups and people whose lifestyle and beliefs do not fit in with the functionalist aims of society, and which may even be dysfunctional, are pervasive, and find their effect in exclusion from housing or at best low-level provision.

One such candidate for exclusion is the young single mother, unless she is rich, famous or both. Young women have become pregnant by males who have then deserted them, or in relationships which have shattered through mutual incompatability, violence, economic and social pressures, or a combination of factors, since the dawn of time. The stigmatisation of young single mothers was such that, in past times, if homeless, they could expect only hostel type accommodation at best, and sometimes confined to mental institutions at worst. Their position regarding housing was made somewhat better under the *Housing (Homeless Persons) Act 1977*, which at least gave homeless single actual and expectant parents priority for accommodation; but what of support needs?

A typical housing response in the 1970s was the 'mother and baby hostel' model, adopted by many local authorities, where sheltered-type housing with a warden was supplied on a group basis, with attendant rules and regulations which may have been somewhat more relaxed than social services provision, but which were in many cases somewhat restrictive. There are relatively few RSLs which cater for this group as such.

The management response which might be thought appropriate would include advice on benefits and employment opportunities, facilitation of child care to enable young lone mothers to find and maintain employment, and housing sufficiently independent to allow customers to enjoy a choice of lifestyle compatible with those who can afford to secure and manage their own accommodation in the market.

Other candidates include young offenders, whose status is determined partly in legal terms, being the subject of legal action including court orders and custodial or non-custodial sentences or lesser sanctions such as anti-social behaviour orders or being party to acceptable behaviour contracts. Assuming that housing providers can grant tenancies, where rents are guaranteed by social services or other official bodies, what are the management implications of providing accommodation? In these cases, great sensitivity is needed to ensure that a full range of support is available to the young offender to help ensure that they do not re-offend due to homelessness, inadequate accommodation, or lack of finance. Here, partnership working between social services, housing providers and probation departments, is essential, with joint care plans and case conferences, in which housing officers as tenancy managers must play a part. The Foyer initiative, which has its roots in France, has been very successful in providing secure accommodation and training and employment, either directly or indirectly, and has been developed mainly by the housing association sector. The idea is basically to provide good quality group accommodation with on-site supervision and training, on a temporary basis, to ensure access to employment, lifestyle and training skills, as a bridge to independent living, and has been around in the UK since the mid-1990s, with a good degree of success.

Street sleepers

What is the image that this term conjures up for you? Is it of a dishevelled character, begging in a shop doorway or near a cashpoint machine asking for spare change, or someone in a sleeping bag under a railway arch, or someone begging on the Tube, indicating that they only want a few pence from everybody to find a place to stay for the night? Or does it bring to mind a sad collection of misfits using a day centre provided by a charity or a council, offering somewhere to get a hot meal and some respite from the streets; or perhaps a soup van in London's Piccadilly Circus? These are poignant images, and somewhat stereotypical, but I believe that the term 'street sleeper' conjured up one or more of these images in your mind.

Street sleepers have been the subject of a large number of initiatives over the last twenty or thirty years. There was the Conservative Government of 1992–96's Rough Sleepers Initiative, which increased funding to organisations like St Mungo's, to enable the conversion of buildings to hostels to get people off the streets and give them basic support to help stay off them. There have been housing corporation-led special grants to RSLs to provide hostel-type or cluster accommodation for such individuals, and there have been somewhat more punitive measures on the part of local authorities to exclude individuals from the vicinity of cashpoints, or attempts to enjoin the police to arrest them under the terms of the Vagrancy Act. There have been private sector attempts to give street sleepers some dignity and an income to help them from vagrancy, like the Big Issue, a magazine sold by homeless people on the streets. But, for all these initiatives, the phenomenon of street sleeping is still evident in our major cities, and beyond.

Street sleepers are neither popular nor deserving in the public eye, and the debate around whether to give support on the streets, or to force them off by inducements or sanctions rages. Those who manage hostels to offer them a degree of support need to be trained in how to deal with the multiple factors which may have led to them being there in the first place, and to help them out of this predicament through training and making opportunities available for rescue. What is needed is an awareness of the agencies which can assist in recovery, such as Department for Works and Pensions-sponsored training and employment programmes and initiative, self-help courses run by charities, and counselling and advisory organisations which can help street sleepers out of alcoholism or drug abuse, if these are present, and bolster self-esteem and social skills which will give them a better chance of returning to mainstream society. The role of housing management is, as elsewhere, facilitatory rather than that of direct provision.

Other victims of societal prejudice

Gypsies and travellers

These are amongst the least popular groups. Gypsies and travellers are often perceived as leeches on society, not paying council taxes or indeed any tax at all, taking over land on a temporary basis and wreaking havoc where official sites are not provided, increasing petty crime rates wherever they go, and pan-handling on every conceivable opportunity, not to mention providing dodgy driveway coverings, building services of dubious repute, and causing a general nuisance. The sign 'No gypsies or travellers' used to be a common one on the doors of public houses throughout the land. It is fair to say that they find difficulty in securing pitches, or permanent housing if they require it.

Housing management operations in these cases follow from duties on local authorities to provide travellers' sites where feasible, although there have been changes in the law which have weakened this requirement over the years. Where such sites are provided, there is frequently a travellers' liaison officer employed by the local authority, although not funded through the HRA as it is a service to the general community. Here the special need may only be for a secure temporary pitch, and the management response to the collection of dues, ensuring that neighbours are not inconvenienced, and that adequate sanitary and other facilities are provided and maintained in good condition.

Others

Students

Students often face an amazing degree of prejudice from would-be landlords or other members of the communities in which they live or need to reside in to be close to their university or college. As one who was privileged enough to attend Cambridge University, accommodation was provided by the college for three years, at sub-market rental rates, and with relatively cheap and good quality food available in the college, not to mention liquid refreshment so necessary to the completion of assignments and dissertations, and to fuel intellectual discussions on the inner meaning of Bob Dylan's lyrics! In many cases, accommodation is provided on some basis or other by the institution, but there is frequently a shortage of available housing in the face of competition from workers and others for relatively cheap rentals in town. The situation has been exacerbated by the decline of the grants system, and top-up tuition fees, which have increasingly placed the onus for payment of tuition and support costs on parents and students, and by the increase in rental costs in some city areas.

In low-demand areas, local authorities have in some cases allocated houses and flats to students on a short-term basis. This has paid dividends in maintaining rent rolls and occupancy, obviating the need for demolition or boarding up, and helping to sustain some form of community, albeit a transitory one, in some areas. In higher-demand areas, this option has not proved possible, and some housing associations have been set up to provide low-cost housing, or have diversified into the student rental market as an arm of their operations.

Housing management options are many and varied in this context. Students are, by and large, articulate and in many cases able and willing to be engaged in the day-to-day management of their accommodation, through management committees and less formal ad hoc decision-making bodies which could be as simple as a house committee to ensure that tasks such as cleaning, refuse disposal, behaviour and commissioning repairs take place. In some cases, local authorities or housing associations have granted leases or licences to student bodies on the basis of self-management, and regular rental payments to the landlord. In other cases, private landlords have signed up with educational establishments on a business tenancy basis, with the university or college agreeing to pay the landlord rent, backed by student financial payments to the institutions, and often entailing self-management.

In these cases, the housing management role may be fairly minimalistic, to ensure that students get access to repair services and financial advice to help pay the rent where necessary, and dealing with major repairs or resident disputes which cannot be resolved by the students themselves. The opportunity for resident participation in housing management is significant in the case of this diverse special consideration group, and the experience of housing management may well lead some to take it up as a career later, if it doesn't put them off.

Victims of domestic violence

Sudden exclusion, or having to leave the matrimonial or shared home as a result of domestic violence or the threat of it is, unfortunately, relatively common, and knows no class division. The law recognises victims of domestic violence as being priority categories for housing assistance, whether or not accompanied by children.

The official homelessness statistics between 1996/7 and 2000/01 show a very slight diminution in homelessness ascribed to the violent breakdown of relationships, averaging around 18 000 per year, which in 2000/01 represented almost one fifth of all officially recorded homelessness. Domestic violence tied with parental evictions is one of the major causes of homelessness (ODPM, 2003b). However, this is a significant underestimation of the actual level of homelessness and housing stress resulting from domestic violence, which is often concealed, or dealt with outside the safety net of official help. In some cases, domestic violence may be a direct result of overcrowded or otherwise inadequate or insecure accommodation. The fact that people often remain in a violent or other unsatisfactory relationship may be a function of inability to find an affordable alternative for themselves or their dependants, leading back to the issue of inadequate supply of social housing.

Recognition of need for emergency accommodation due to domestic violence is one thing, but feasible and supportive housing management strategy is another. The victim of domestic violence may need immediate medical attention, and emotional support, best provided by specialists or friends. Housing departments can provide immediate contact with appropriate medical and support services, as well as emergency accommodation, either directly or through other bodies such as Women's Aid, which is often the recipient of General Fund assistance to help with running costs, and which can give help with legal actions, social and domestic social support, as well as providing a relatively secure refuge from the perpetrator or his associates.

In such cases, trying to maintain the confidentiality of location is essential, as well as liaison with the police. In the longer term, councils can often assist by enabling victims to determine joint tenancies where there is the possibility of effective injunctions aimed at preventing approaches or re-entry to the home by the perpetrator, or by facilitating transfers to another authority or area to help minimise the threat of further violence or intimidation.

Housing advisory officers can, as well as dealing effectively with applications under the *Homelessness Act 2002*, ensure that legal advice and assistance is given to the victim, and require a good knowledge of matrimonial law in respect of property, on a cross-tenure basis.

The management of temporary accommodation

In 2004, there were approximately 60 000 households accepted as homeless and in priority need in some form of temporary accommodation in London alone, and around 90 000 nationally (31 March 2004), rising from around 78 000 at the end of September 2001.

Temporary accommodation use has risen over the past ten years, even though there has been a less significant increase in homelessness acceptances nationally. The reason is that turnover – the availability of new and relets in the social housing sector – has reduced in many areas. Crudely, temporary accommodation is silting up, as waits for social housing get longer.

For an official account of the causes of homelessness, and various stopgap temporary housing solutions employed up to 2002 to stem bed and breakfast usage, see *More than a Roof* (ODPM, 2003b), which predictably slants towards a preventive approach, and does not relate the phenomenon to supply shortage to the extent that more critical commentators have.

Since the limitation of bed and breakfast hotel usage to discharge interim duties in respect of homeless households with children, except on a time-limited basis in emergency, since April 2004, the predominant form of temporary accommodation has been in the form of housing secured through the private sector on lease basis, either directly by the local authority or through an agent, typically a housing association or dedicated company.

The former method is known as private sector leasing (PSL) and the latter – where the housing association is the management agent for the owner – as housing associations as managing agents schemes (HAMA); or where the housing association takes a lease from an owner, and accepts nominations from the local authority, as housing association leasing schemes (HALS).

Typically, these schemes involve offering housing on a temporary basis, pending the applicant finding accommodation for themselves, or through the local authority or a housing association. Where the housing is supplied by direct leasing from an owner by a council or managed under HAMA, the tenancy is typically of the non-secure form, and where a housing association leases, the occupant is generally an assured shorthold. This was true at the time of writing, although the correct term may well now be a Type Two tenancy, due to tenure reforms in the mid-2000s. The term 'temporary' needs qualification – stays are often two or three years, depending on the availability of permanent housing, determined by the level of ADP development in the case of RSLs, or turnover in the permanent stock in the case of councils.

There are a number of housing management implications for temporary accommodation. If the property is leased from a private owner, the terms of the lease may entail regular repairs and inspections, which means close liaison between the tenant and the manager to gain access in a non-intrusive manner.

The property may be located in an area where there is little renting, and adjacent owners may fear for their property values if they know that the occupants are 'homeless' or dependent on welfare benefits, which may or may not be the case, and so the housing manager may well have to step in to stop harassment.

The accommodation may well be superior to that which the council or housing association can hope to provide, so there is the task of fully explaining tenancy rights and limitations at the outset of the let, as well as the issue of trying to ensure vacation when a permanent offer is made. Choice-based lettings schemes, which may assign a large number of 'trump card' points to households in temporary housing where the head lease is up for termination, or surrender, and where there is no prospect of lease renewal, may help to soften the blow by presenting a range of alternative social housing lets at affordable rates.

Then there is the issue of affordability. The financing of such schemes varies, but in many cases, the market rent payable to the private owner is backed by a rent to the occupant which is in many cases at least equivalent to the lease fee, within the limits imposed by the housing benefit (HB) scheme. Sometimes the occupant rent incorporates a contribution towards those costs of managing and maintaining the property which are down to the leasee. The housing benefit payment is backed in turn by housing benefit paid to the leasee or the occupant, or by housing benefit subsidy paid to the General Fund by the DWP to offset the loss to the authority of rent rebating.

In these cases, a large part of the housing management task may well be to ensure that the HB claim is made swiftly, and that HB is paid in a correct and timely fashion, to minimise the gap between payment to the council or RSL, and onward payment to the owner, dependent on the frequency of the head-lease payment negotiated. Arrears resulting from delayed HB payment, which in some circumstances will be irrecoverable, will threaten the viability of schemes.

In cases where the occupant is in work, and not eligible for HB which would fully cover the rent charged, the management task may involve advice on budgeting as well as liaison with DWP or the council to ensure that tax credit or similar support is claimed, to ensure that the occupant stands some chance of affording the rent. Alternatively, the task may involve securing a move to more affordable housing as soon as possible, whilst minimising the stress of moving.

This is in addition to all the standard housing management tasks expected; not an easy role, but one which will remain as long as rates of new permanent housing provision are less than required. We could be in for the long haul here.

Involving the customer

Social housing tenants and those applying for housing are consumers of a variety of services as already identified, but in what sense are they customers? This has been the subject of academic debate for some years, much of it rather sterile. A consumer is someone who receives a good or service. A customer is someone who gives custom, normally interpreted as a consideration, in return for a good or service, and enjoys a bundle of rights in respect of that delivery in consequence of having entered into a contract with the supplier. Social housing tenants are customers in the sense that they have entered into a contract with their landlord in return for a consideration (rent), and the contract incidentally requires that the customer abides by contractual terms, including repair obligations, duties to ensure the quiet enjoyment of neighbouring property, desisting from anti-social behaviour, and so forth.

It is true that the payment of rent may not come from the tenant's own resources, as it may come wholly from housing benefit paid on behalf of the state, or not be demanded at all, and be covered by a rent rebate. It should be noted that the latter is a form of housing benefit, and housing benefit is a personal benefit, which becomes part of the personal resources of the tenant, just as much as earned wages or other forms of income support are. It is the tenant's money, which may or may not be used to fulfil the terms of the rental contract.

There is some debate as to whether some supported housing tenants, where responsibility for contractual dealings may be taken on by another person or agency, are truly customers as they may be incapable of entering into or servicing a contract through not being fully aware of the term of the contract, and that therefore they may not have the same range of rights as other social housing consumers. However, the resolution of this debate is beyond the scope of this book, and is identified purely for the reader to reason through. The bottom line is that most social housing tenants are customers in the formal sense of the word, and therefore have customer rights, as well as the right to be treated as customers.

Whether applicants for social housing tenancies are customers is a moot point. At the point of application, they are seeking a good or service and have not, by definition, obtained what they are seeking: they are potential social housing tenants. On the other hand, they are customers of council or housing association applications services, and the consideration is the currency of housing need, expressed formally in many cases by the number of points allocated to the application.

Customers of services provided in the private sector are getting ever more demanding, even in England, if the welter of television consumer programmes is to be believed. The National Consumers' Association is a highly visible organisation, and the growth of the consultation culture is proof enough of the increasing assertiveness of those who believe they have not received value for money, and who wish to complain.

The same is true in the social rented sector. Largely in response to consumer group pressure, councils now have to publish performance indicator information against publicly set targets on all areas of their delivery, ranging from education through to waste collection, and housing has not escaped. With the *1989 Local Government and Housing Act* came the requirement to publish an annual report to tenants and housing performance indicators on key areas including rent collection levels, void times and numbers of people housed by ethnic category. These indicators have evolved through the Best Value regime and beyond. The 1990s saw John Major's Citizens Charter, which spelt out the level of services which people could expect from their councils and public utilities, and tenants are deluged with information from all sides on matters ranging from large-scale voluntary transfer, private finance initiatives, Arm's Length Management Organisation proposal, rent restructuring and other fascinating topics, translated into all community languages, and available as downloads on dedicated websites.

Both ODPM and the Housing Corporation issue frequent guidance to councils and associations on how to involve tenants and leaseholders in the management, maintenance and strategy of their organisations. A recent contribution to the vast literature, of practical use to housing association managers in particular, is the Audit Commission's report and associated management handbook. The central message is that, as well as being self-justifying, resident involvement can actually lead to more cost-effective housing management, and more responsive and accountable governance (Audit Commission, 2004).

Sometimes, information is all that customers require. A clear manual for a car would be good, and would help keep the service costs down and maybe help ensure competent if not adventurous driving, but I may not be tempted to do my own servicing as a result of reading it. One does not follow naturally from the other. On the other hand, if I wanted to get stuck in, I could do so, if the manual was clear and comprehensive enough, and I had a modicum of training, and the time to do it.

Information is, at its most effective, a two-way flow. Housing organisations do well to conduct satisfaction surveys from time to time, and to request feedback on the quality of repairs, so that they can continuously improve. They also do well to ask whether tenants want to get involved in the management or scrutiny of the services they receive. Once you start opening the information door, and invite comments, you are bound at some stage to get a response to the effect that you don't know what you're doing, or that the customer could do a better job themselves. This is precisely what has happened in the housing field. The format of tenant surveys is, to a certain extent, laid down by regulatory bodies such as the ODPM (for councils) and the Housing Corporation (for registered housing associations), and the involvement question has been a requirement for some years now. It has to a certain extent stimulated the movement from information to tenants to tenant involvement in housing strategy and management.

Another stimulus to more general involvement is long history of tenant involvement in the voluntary sector. Housing co-operatives, where tenants call the management shots by comprising the board and are directly responsible for the fortunes of their organisation, have been around for a long time.

Self-management

The most significant shot in the arm for the tenant management movement came in 1993, as a result of the *1993 Leasehold Reform, Housing and Urban Development Act*, which introduced the Right to Manage. This gave secure tenants of local authorities the right to manage their own properties, as long as there were more 25 or more homes in the package. It introduced the term 'tenant management organisation' (TMO) into the housing vocabulary.

The governing rules for setting up a TMO were set out in the *Housing (Right to Manage) Regulations 1994* (DOE/Welsh Office, 1994). Under the Right to Manage (RTM) , tenants can take on a variety of housing management functions, ranging from rent collection, repairs, allocations and lettings, and leasehold management; to involvement in major repairs and improvements planning, under contract to the council, and supported by an allowance based on what the council would have spent doing the same functions, plus an amount to cover committee training, and staffing.

The Modular Management Agreement (MMA), introduced in 1994, offers a variety of different management options, and

ODPM provides money in the form of a grant to support the agreement negotiation process, through *Section 16* of the *1986 Housing and Planning Act.* Negotiations are assisted by so-called 'section 16 agents', independent tenant advisory organisations like the Tenant Participation Advisory Service (TPAS) or Priority Estates Project (PEP), who also offer training and other support to ensure that the tenants and residents' associations which initiate the RTM process. At the time of writing, the RTM agreement is being revised by ODPM to take account of important changes in the law on leasehold management and equal opportunities legislation, and developments under the Best Value regime, principally more sophisticated performance indicators, amongst other changes which have taken place since 1994.

The TMO can only take over management if it secures a majority vote of secure tenants in the proposed management area by ballot, usually conducted by an independent organisation such as the Electoral Reform Services.

There is little doubt that tenant management organisations have succeeded in empowering tenants, and have also had collateral benefits such as management and administrative skills training. The diversity of approaches to TMOs, and the variety of benefits that they have brought, are highlighted in several reports, amongst them *Tenants Managing* (ODPM/Cairncross *et al.*, 2002), and more generally in the Chartered Institute of Housing's report, *Resident Involvement and Community Action* (CIH, 1998).

There is no analogous right for housing association tenants, but the Housing Corporation has encouraged housing associations to extend housing management involvement to its customers over the years. It should be remembered that many housing associations grew out of the co-operative movement, which had resident management and control as its central philosophy (Housing Corporation, 2003a, 2004a).

The Right to Manage built on the great success enjoyed by Tenant Management Co-operatives from the mid-1970s onwards, and was promoted partly through the work of Anne Power, who proposed management solutions to turn unpopular estates round rather than the rather crude estate remodelling and walkway-demolition moves advocated by Alice Coleman in her *Utopia on Trial* (1986), although estates which have been altered in this way have often been the subject of the RTM as well.

The process of negotiation with a local authority can be a long one, determined both by the immense work needed to ensure that every option in the MMA is carefully considered, the intricacy of working out allowances, and the legal detail entailed in ensuring that the options hang together. It is also essential that tenants receive capacity training to ensure that they can provide an excellent management service, albeit in most cases with the help of a paid officer or management team.

The process of capacity building does not stop when the agreement is signed – there is need for constant support and updating, as well as ensuring that the management committee does not atrophy, and that the door is open for new board members from all parts of the community, and that the committee publicises its work and reports back on its performance to all stakeholders.

There are many other instances of participation and involvement, and it must be remembered that there is a ladder of participation, ranging from information giving, through estate committees, which scrutinise the council's housing management at local level, joint management boards, where operations are overseen by council members and residents, right the way through to TMOs. There is no 'one size fits all' approach, nor should there be, and there is plenty of scope for enhancement, given training and growing familiarity through practice.

It is, of course, not necessary for tenants and leaseholders to take over the management of their estates and properties to have influence over their landlords; in fact, the whole idea of involvement at all levels is ingrained in current housing policy. In the late 1990s, local authorities were enjoined by government to negotiate Tenant Participation Compacts with their rent and service-charge paying residents, with the aim of ensuring that they are involved and informed at the level they wish to be, and that the organisations which represent them are truly inclusive of the communities which feed them. These compacts do not have the status of legal contracts, but councils' housing performance is judged largely on the way they measure up to the standards negotiated. In many cases, compacts have led to devolution of budgets, the formation of area housing panels or committees, and the renewal of tenants' and residents' organisations, and their council-wide federations.

It would be a foolish council which did not take note of its tenants and leaseholders' aspirations.

Many key government initiatives over the past few years have relied explicitly on tenant support through the ballot box, and local administrations have fallen as a result of 'no' votes arising in many cases through fundamental distrust of those initiating the moves, aided and abetted by radical organisations such as Defend Council Housing, which may have sometimes conflated alternative public realm management approaches with privatisation.

The key 'Options' under Options Appraisals of the first decade of the millennium, such as ALMOs, PFI and Large Scale, Small Scale or Partial Transfer have explicitly or implicitly relied upon positive support from tenants, although a ballot is only required in the case of the transfer options, at least where the properties are occupied. A brief guide to tenant involvement was provided by ODPM in 2003 (ODPM, 2003c). Councils were told that they could not continue to directly manage their stock if they had no prospect of bringing their homes up to the Decent Homes Standard by 2010, and that they had to choose one of the 'options' after appraising their likely effectiveness and consulting with tenants. None of the options has proved easy to implement, partly because of tenant scepticism over whether the new organisations would be any more effective than the democratically elected council bodies they were to replace, and despite assurances of tenant board membership – effectively resident power-sharing which had been notably absent in many authorities beforehand.

Tenants have shown that they cannot be made over with simplistic presentations or wild promises unsupported with coherent, well-founded and comprehensible business plans, even where it can be readily demonstrated that councils lack resources to continue to provide a decent service. The failure of the Tenants' Choice initiative following the *1988 Housing Act* is a poignant reminder of this fact. Local authorities have frequently not made it easy for trust to develop. Years of traditionalistic, paternalistic housing management cannot be replaced by a sudden change of heart. The distancing of local decision making from ordinary councillors and their electorate through the Mayor and Cabinet model of local governance, and the incomprehensibility of much town hall bureaucracy, for all the glitzy websites and lip-service to the customer-care ethos which has come on since the Citizens' Charter days of the early 1990s, have all played their part in alienating tenants and leaseholders. Radical neglect and appalling repairs and maintenance performance on the part of many councils, aided by a lack of central resource help and plain mismanagement in many case, despite the rigours of Compulsory Competitive Tendering (CCT) and Best Value, has not helped municipal credibility and has made the task of presenting and implementing valid management re-engineering and ownership approaches all the more harder.

Neither has the housing association movement helped itself, or local authorities, in these respects. Housing associations often have unelected committees and self-appointed boards, offer antiquated and paternalistic constitutions, and often fail to give support for genuine tenant management and involvement, despite Housing Corporation and industry-body enjoinment. The very poor image that many associations offer has not helped the drive towards disposal to them, or newly created organisations, dependent on the consent of paying customers, who often mis-perceive them as private companies intent on charging as much as they can get away with for their assured tenancies, or pushing home ownership at the expense of investment in housing for those who cannot contemplate owner-occupation.

It is clear, for all the talk of participation and compacts, of self-management, local budgets and ownership, that there is still a very long way to go along the road of genuine and wholehearted participation.

A tenant satisfaction questionnaire or repairs response card is not a manifesto for encouraging tenants to take control of their housing destinies where this is wanted.

Repairs and maintenance

Most tenant surveys, across the social rented sector, indicate that the state of repair of the home is a primary concern, along with neighbourhoods free from harassment, crime and violence. Government policies in the early twenty-first century have been driven by the decent homes and investment agenda, and have encouraged councils and housing associations to become more effective in maintaining the homes they own and have stewardship over. Every council which owns stock must have a Housing Revenue Account Business Plan which must aim to be Fit for Purpose; and one of the main criterion for fitness is that there should be a sound repairs and maintenance strategy. Performance indicators on repairs standards abound, and just about every tenant satisfaction survey has repairs as a focus. TMOs have been formed on the basis that the council has done a lousy job, and in the belief that tenants can do it better with the same level of financial resource. Stock has been transferred or been made subject to Private Finance Initiative 30-year improvement programmes on the basis that councils have been unable to find the right level of resources to provide a decent repairs service, let alone major improvements. The virtual elimination of new council housing development since the late 1980s, and the disappointingly low level of housing association starts in areas of great need, have increased the importance of making ageing social homes last by making sure that they are repaired and maintained properly, against a background of dwindling cash resources and inadequate central assistance. No wonder many tenants have sought a kind of refuge, often misjudged, in the Right to Buy.

Repairs are responsive fixing jobs, for example when a washer needs replacing or a broken window needs replacing. Sometimes, these jobs are known as 'responsive repairs' to distinguish them from planned works. Maintenance can either be planned or cyclical. Planned maintenance is where a landlord has programmed works on a street, estate, or block, and where they are scheduled on an ordered basis. For example, roof maintenance may be required on an estate, and the contractor is told to commence on 1 April at Block A, and work their way round to Block C, and progress is monitored on a quantitative and qualitative basis, with time and quality checks carried out by the client or operatives' management.

Cyclical maintenance is where maintenance is carried out on a regular basis – on a cycle – say, on a yearly basis, for example in the case of boiler maintenance. Another example is a cyclical painting programme, where doors and window-frames may be painted every five to seven years. All of this can be distinguished from improvement works, which involves enhancing the value of the asset, for example by bringing it up to a decent state rather than just keeping it that way.

There is a clear link between repairs, maintenance and improvement: if a landlord fails to carry out repairs, or to undertake cyclical or planned maintenance, the properties will soon become defective, and may even become unfit for human habitation. If the properties are then abandoned, they will need to be improved – re-roofed, new kitchen fitments installed, redecorated and generally modernised – before they can be relet. A landlord may decide to improve properties, for example by replacing out-of-date but sound kitchens and bathrooms with layout and equipment conforming to modern standards and expectations. Another example is where the property is extended to provide more spacious or convenient accommodation.

Repairs and maintenance are generally paid for through revenue funding, through the Housing Revenue Account of councils, or Property Revenue Account of housing associations. Some of the expenditure comes through rent income, or through a mixture of central government and locally generated housing income. In the case of housing associations, there is in most cases no central revenue subsidy from the Housing Corporation, and repairs and maintenance must be met through rents. In some cases, where these obligations cannot be met from these sources, landlords have 'capitalised' repairs, that is, paid for them by using loan funds or cash held on deposit, sometimes raised through the sale of assets.

The major repairs programmes – effectively very large-scale maintenance programmes – of local authorities were, before 2001, largely funded through the Housing Investment Programme money raised by way of loan and paid back over 30 years.

Since 2001, councils entitled to HRA subsidy have received a Major Repairs Allowance, based on the number of property 'archetypes' – property types differentiated by bedroom number, traditional or non-traditional methods of construction, type (flats or houses), age, and number of storeys up – in the stock, and aimed at helping councils to maintain their stock at existing standards rather than enhancing them. The move was partly in response to the evident repairs and maintenance backlog which had built up since at least the 1980s due to a combination of too low management and maintenance allowances, unwillingness to raise rents to levels to meet the significant costs entailed, and restrictions preventing the cross subsidisation of HRAs from General Funds (see DETR, 2000). The number of archetypes – in 2004, 13 – has been shifted from time to time, as has the level of MRA: and it has been referred to as the 'revenue-isation of capital'. This is in addition to allowances paid to support responsive repairs and non-Major Repairs cyclical and planned maintenance through the Management and Maintenance Allowances to HRAs, where they qualify.

Housing associations frequently use cash built up through rent surpluses and held as a repairs and maintenance sinking fund to undertake programmed repairs and maintenance, but have significant freedoms to capitalise them as well, although backlogs have built up in the voluntary sector as well as a result of a combination of mismanagement (in some cases), lack of judicious planning for repairs eventualities when properties have been built, and miscalculation of rental levels.

Neither councils nor housing associations can raise rent levels simply to pay for more or a higher standard of repairs, or to backtrack and meet backlog needs, due to the constraints imposed by rent restructuring, but more of this in the next chapter. Suffice it to say that a major preoccupation of housing managers is to find ways of making repairs services and materials procurement more cost-effective, whilst attempting to ensure that tenants and leaseholders abide by their contractual and statutory duties, outlined in the lease or tenancy agreement and handbook.

Tenants of social landlords do have repairs obligations. They must replace or repair items damaged by themselves, minor fixtures and fittings, and most things that they have installed. Leaseholders are generally responsible for repairing most fixtures and fittings within their dwelling, but not common parts like lifts, or structural items, for which a service charge is payable. In the case of tenants, if the landlord fails to action a repair, it is ultimately possible for them to order the repair themselves and bill the council, under the Right to Repair, after some rigmarole involved in trying to get the landlord to do it. If the landlord consistently fails to observe their repairing obligations, tenants and leaseholders can go to court to force their hand, through a disrepair action, or to get compensation.

In some councils in the 1990s, the legal cost of disrepair actions was in some cases equal to or higher than the repairs budget itself.

Inadequate repairs services have often motivated tenants to take over housing management themselves, in the form of TMOs or joint management boards, have given something for Tenants and Residents Associations and Leaseholder Associations to focus on, and have even motivated them to choose Large Scale Voluntary Transfer on the basis of councils' frank admittance that they could not hope to provide a decent repairs service, let alone meet the backlog of repairs and maintenance, let alone undertake improvements to the standards that tenants demand.

The search for economy, efficiency and effectiveness in the repairs service was a major stimulus to Compulsory Competitive Tender (CCT) regime of the early 1990s, following on from the CCT of rubbish collection and other key council services in the 1980s. CCT involved compelling councils to invite tenders from other, non-council providers, to undertake housing management and maintenance services. The winner would be decided on the basis of price and quality. The choice of contractor was exempted from the requirement to consult tenants on matters of major housing management change by a change to *Section 27* of the *1985 Housing Act*. For all the bluster, housing CCT take-up was remarkably low, compared to the virtual privatisation of rubbish collection, public transport, school meals and street-sweeping services, largely because the margins made no commercial sense, although some companies were founded on the prospect, and there were several management buy-outs and CCTs involving housing associations. One local authority – the Royal borough of Kensington and Chelsea – even formed a borough-wide TMO (TMOs were exempt, as they were already a management organisation under contract to a local authority) perhaps in order to avoid the compulsion to go down the CCT route, although it wasn't badged in this way at the time!

Some councils and many housing associations shed their repairs and maintenance Direct Labour Organisations (DLOs) in any case, to try to make economies or in some cases bat off the problems of labour relations which had been a thorn in their sides in the 1970s and 1980s. Unfortunately, having done so, they often found it harder to control the quality of works than before, and found themselves in trouble when they determined the contracts of under- or non-performing contractors, only to find that no-one else wanted the job.

With the failure of housing CCT, and a new administration, a new regime was put in place – that of Best Value, where, instead of having to tender out housing management and maintenance, councils were compelled to compare their services with other providers in quality and cost terms, and thoroughly consult residents before making decisions, without being forced to contract services out. The new regime came into force from 1997, and is still more or less in place. Housing Associations have never been subject to these strictures, but are forced to take great care in selecting contractors due to the rigour of the performance indicators imposed by the Housing Corporation, and the threat of deregistration, imposition of board members, or restrictions in access to the Approved Development Programme, not to mention pressure from tenants and leaseholders.

Outsourcing can be a way of stimulating the local economy, and enabling unemployed tenants and their households into work. A number of Resident Service Organisations, where local people and especially tenants are employed, have been established by local authorities and housing associations throughout the European Union as part of economic regeneration initiatives. They have been given training and set-up grants, sourced from a variety of funds ranging from individual or cross-borough grants, national funds, and European Economic Development Grant pots, to empower residents to compete in the marketplace, and many have won repairs contracts on their own terms. Such schemes are complementary to modern apprenticeship initiatives, and programmes established by Technical Employment Councils and Neighbourhood Regeneration bodies. Arguably, involvement in such schemes gives residents greater buy-in to the issues in their areas than otherwise, as well as cash in their pockets, to help lift them out of poverty, and is wholly in line with the choice and responsibility agenda of New Labour's approach to welfare benefits in the early twentieth century, for all its evident faults.

Local further education colleges run courses on repairs and maintenance for residents wishing to establish Resident Service Organisations (RSOs) and other similar bodies. Given the acute shortage of builders, carpenters, plumbers and maintenance workers, especially in London and the South East, the future of such organisations seems assured, and such work looks to be a good bet for income accumulation for those with the initiative and foresight to take these trades up.

Quality matters

Local authorities are now judged for the quality of a variety of services, including management and maintenance, through the Comprehensive Performance Assessment regime simplest, it is a methodology for moving from the present situation, to the specified goals, aims and objectives of an organisation. It should be said that even tentative promises of financial support usually rely upon the existence of a well-structured plan setting out the scale of likely demand (the size of the market), its dynamics (will the demand last long enough for a business to meet it in the short, medium and long terms?), and the scale of the competition and likelihood of market-saturation as other companies are formed at or around the time which the enterprise in question is set up. Financiers will also expect to be certain that the entrepreneur has priced its products or services competitively, that the quality of these is likely to attract sufficient custom, and that there is a credible advertising and marketing plan in existence, at least in embryo form. Other considerations include a plan for premises and capital equipment acquisition and servicing, tax considerations, and even a viable exit strategy should the enterprise fail. The issue of number of staff ('the establishment') and remuneration is also relevant. In short, the financier will expect the entrepreneur to demonstrate that he or she has a plan to meet goals in a costed fashion, before the financial body will consider risking cash.

In short, business planning involves the following steps:

1. Specification of organisational goals – the Mission or Vision Statement.
2. Specification of aims (the goal broken down into its elements), and objectives (the aims broken into a series of timed stages).
3. Identification of the gap between where the organisation is now, and where it wants to be (at a given time).
4. SWOT analysis, i.e. setting out the strengths, weaknesses, opportunities and threats – both in terms of the internal (organisational) and external (market, existing suppliers, economic) environment.
5. Options planning – designing various strategies for getting from present position to goal, identifying the financial and personnel resources necessary to achieve goals.

operated by the Audit Commission on behalf of the ODPM. A one- or no-star rating on the basis of a poor repairs service can make the difference between being able to participate in the new financial freedoms – to borrow privately, and have greater freedom in raising and using money on behalf of communities – and having to do things in the same old way under continuing restriction. Housing services are inspected by the Housing Inspectorate, and again, a poor repairs rating can have implications for being able to go for ALMO status, or at least receive financial support for options calculated to improve the lot of tenants. Housing associations receive scrutiny by the Housing Corporation, and its agent in this respect, the Audit Commission as part of the former's regulatory remit (Housing Corporation, 2004b).

Tenants are often asked to feed back on the quality of repairs services, whether contracted out or provided in-house, by filling in repairs satisfaction cards whenever a job is done, and through tenant satisfaction surveys. Annual Reports to Tenants produced by both housing associations and councils, always report back on repairs performance, focusing on numbers and percentage of repairs done on time in emergency, urgent, routine and planned/cyclical categories, quality satisfaction data, and other vital statistics which can be compared year-on-year, that is, when the basis for collection does not change.

A repairs service for the future

Repairs are a nuisance: householders do not want the inconvenience of faulty equipment, or to suffer the risk to security or health through neglect of repairs. They want a timely, responsive service, provided at their convenience rather than that determined by the needs of some bureaucratic system. They want operatives who are identifiable, polite, knowledgeable, efficient and effective. They want the repairs to last, and maintenance programmes to keep their homes in a decent condition throughout their tenancy. Is the aspiration to live in a safe, secure, warm and well maintained environment too much to ask for?

In a country where most people have access to a phone, and many have an internet connection, an efficient and effective repairs system is one where the customer can contact the landlord as soon as the repair need is identified, through a variety of media, and can expect to be put through to the right person straight away (perhaps via a call centre), or log the repair themselves electronically. It is customer-responsive to let know when the repair will be done in terms of its priority ordering, or to give advice on which company or companies to contact if the repair cannot be carried out within the specified timescale, or if it is not the landlord's responsibility. Repairs should be carried out at a time convenient to the customer, with the minimum of fuss and delay, and inspected on a sample basis to ensure quality, backed up by customer feedback on a repair-by-repair basis. Complaints should be dealt with quickly, fairly and effectively, with compensation available at appropriate levels by way of apology. An excellent checklist for a good repairs service in the social housing sector is to be found in the Chartered Institute of Housing's good practice manual (CIH, 2001).

Hopefully, this is not just futurology, but a legitimate and worthy goal for social landlords to aim for. All that is lacking in many cases is the financial resources to do so, and it is to this subject that we turn in the next chapter.

Some organisational management issues

Housing organisations are outfits with aims, objectives and plans to achieve them; and they do so through doing jobs and raising finance to back their operations. They would not be able to do anything without staff: and personnel are the most valuable resource of the social housing organisation. If it is anything, housing is a people business – social housing bodies deal with the hopes as well as the homes of their customers, and influence their life-chances and lifestyles far beyond the provision of bricks and mortar.

How, then, are social organisations staffed, and what are the personnel and management structures like? How are competent personnel recruited and retained, trained and nurtured, rewarded and disciplined? And how are organisations structured, in terms of hierarchy and spatially, to

meet customer needs, and the requirements of legislation and custom? It is vitally important to understand that the strength or weakness of any social or commercial organisation lies mainly in the attitudes and attributes of the staff who make them up. Far too often, staff needs, and the structure and command chains of organisations, are overlooked, and the human dimension subsumed to one of 'human resources' where job descriptions and roles are held to be paramount, and the celebration of ambition, positive attitude and personality in adding value to essential services is marginalised, often with disastrous labour relations and delivery consequences.

This part will cover retention and recruitment, hierarchical structures, remuneration, business planning and staff involvement in helping to determine the aims and objectives of organisations and workplace practices.

Recruitment, job design and business planning

Recruitment to the housing profession is achieved in a number of ways, ranging from the traditional advertisement, interview and/or test route to headhunting for senior posts by human resource consultancies. Prior to advertisement, it is essential to assess whether the post, which may have arisen through staff turnover, is actually necessary to help meet organisational aims, and if so, whether it is properly specified in relation to those aims. Job design, and the compilation of essential and desirable personnel criteria which will be used as selectors, should relate to the business plan, and be considered in relation to other posts in the organisation (see Green, 2004 for good practice guidance, which reinforces these points).

It is only when an organisation is set up, or is to undergo substantive revision, that job descriptions and specifications are usually reviewed or designed up, and then it is customary to use models from other bodies with analogous needs; and there are plenty of those. So what are the links between job design and the business planning process?

Business planning is at the heart of effective organisational specification, re-engineering and job specification. At its

This is the part of the business plan which potential backers are most interested in.

6. Qualified decision, based on all of the above.

This is a distillation of thinking on what constitutes a business plan, and the reader may find Smith (1997, Chapter 4) a reference useful as background reading (see also Blundell and Murdock, 1997).

Having constructed a business plan, it should be possible to determine the organisational roles necessary to achieve the aims and objectives set out. This can be exemplified in Table 2.3.

This is a vastly simplified attempt to set staff requirements against objectives, but how many times does this form of planning actually happen, even in a rudimentary form? In most cases, the organisation already exists, and it is a matter of periodically reviewing staff roles against organisational requirements at points of change, or when there is an obvious mismatch between requirements and staffing. In very many cases, organisational aims and objectives seem to be fitted to the staff who happen to be around at any one time, or the existing roles have an importance and sanctity beyond any organisational change imperative. It is almost impossible to start with a blank sheet of paper when trying to fit roles to organisational needs, unless the organisation itself is new. But it is important to review requirements from time to time, to ensure best use of staff skills to meet objectives, and business planning is an approach which has yielded fruit. A distinctive feature of properly conducted business planning is the involvement of staff at all levels in determining strategies to grow the business or cope with change, especially where the organisation is multi-layered, with sections, departments or divisions which are charged with delivering specific objectives, as is the case in most social housing organisations.

On the assumption that the roles in an organisation have been properly specified in relation to needs, it is now necessary to ensure that job descriptions and personnel specifications reflect those roles. The job description outlines the duties of the post, and level of responsibility, as well as

Table 2.3 Social Housing Organisation X – roles against aims and objectives (extract)

Aim: To provide an effective housing management service in XYZ Council	
Objectives	Staff needed
Provide effective tenancy management service	Personnel with good knowledge of tenancy law and experience of dealing with social housing tenants
Provide an effective customer communications service	Personnel with proven ability to deal courteously and knowledgeably with social housing customers, using all available media e.g. telephone, face-to-face, electronic communication
Provide effective administration of records on tenancy details, requests (e.g. for transfers and exchanges) housing benefit status, dealings with other council departments, etc.	Personnel with proven administrative capability in housing or other related fields, able to use computer and hard-copy databases
Provide effective financial management of the operation	Qualified accountancy personnel to oversee financial operation, and staff with abilities and experience in financial records management and in preparation of management accounts and balance sheets, income and expenditure accounts, and other relevant financial statements.

who the postholder reports to, and may change with organisational modification. The personnel specification identifies the skills and knowledge attributes required to fulfil the objectives of the post, and are often divided into 'essential' and 'desirable'. For example, in the case of a rent collector, an essential attribute may be the ability to understand and communicate the contents of a rent account to a customer – the role could not be performed by someone who could not do this, even if training is required. A desirable attribute may be the possession of a clean driving licence: it may not be essential, because the operative may use a cycle or public transport. These criteria form the basis for drawing up the advertisement, assessing the remuneration and some of the post-specific terms and conditions, and are used to shortlist candidates, and as a framework for questions and decision at interview and test stage, and ultimately, as a method of selection on the basis of how well candidates measure up to the specification.

Legal matters

In drawing up personnel specifications, organisations have to have regard to employment and related legislation. This is a shifting field, due to modifications in equal opportunities legislation, but the principle is that selection should be on the basis of the requirements of the role rather than on any purely incidental factors, such as race, ethnicity, age, sex or religion, unless any of these are a Genuine Occupational Qualification. This is clearly important on its own terms, but also to avoid expensive litigation, and to meet the strictures of regulators and monitoring bodies. A regular health check on whether equal opportunities and employment legislation in general is being observed in recruitment matters is prudent, as is competent training to managers and others charged with hiring and firing (Green, 2004).

Staff retention

Retaining good staff in social housing organisations has often proved difficult. One of the reasons is that there is considerable choice, given the number of councils and housing associations in close proximity to one another, within reasonable travelling distance of job seekers. Dissatisfied staff can always seek opportunities elsewhere, especially if there is a shortage of staff in relation to available posts, and the potential for, or actual, leakage to the private or other parts of the public sector, where skills are transferable.

This is especially, but not exclusively, the case for finance and information and communications technology staff. Internal promotion opportunities do not always exist, and the tradition of moving up when the manager leaves has long gone, even if it ever existed on a large scale.

Remuneration may also be another cause of turnover. Local authority and housing association jobs, especially at lower grades, are often uncompetitive with similarly skilled roles in the private sector, and there is little uniformity on remuneration for similar jobs within and between the council and housing association sector in any case. The keyworker housing crisis identified in the late 1990s and early 2000s, where essential public sector workers found it difficult to find affordable housing in economic hot-spots, with implications for the social infrastructure of such areas, illustrated the relatively unfavourable pay situation in the public sector, and helps to explain the continued shortage of teachers and hospital staff in many such areas.

Terms and conditions apart from pay may be another factor explaining turnover rates, good or bad. Traditionally, local authority workers have enjoyed relatively favourable terms and conditions, with office workers working basic 35- to 37-hour weeks, relatively generous leave well in excess of the statutory minimum, good training facilities, an excellent final salary pension scheme with a guaranteed lump sum and pension based on years of service and final salary, and with the services of trade union representation.

Many of these features are still present, and should not be under-valued, but there has been change in all of these aspects. There is no such thing as a job for life, with organisations downsizing, with staff being transferred initially under the TUPE (Transfer of Undertakings in Employment) regulations to new-style bodies such as ALMOs or LSVT associations, where the new body changes its approach to employment when original staff leave. Increasingly, the advantage-gap is decreasing; and with the constant threat to the final salary pension scheme, and its virtual absence in the housing association sector, there are progressively fewer advantages to working in the public or voluntary sector compared to similar private sector roles. Nor do such organisations have the financial muscle to compete effectively with the private sector, due to limitations of local and central revenue raising ability, and competing demands on budgets.

What, then, can be done to retain good and committed staff in the public sector, and specifically, in housing management?

First, the field may require an image makeover. The management and maintenance of social housing may not be the most glamorous occupation around, even in the public sector. With demographic changes implying a reduction in the number of economically active people due to the effect of reducing birth rates, unless the situation changes due to an influx of economically active households from the EU accession countries and elsewhere, or through changes in the age of pension eligibility, housing organisations will be competing increasingly for a shrinking pool of potential employees. The central problem in the future will not be selecting out unsuitable applicants from a vast postbag of applications, but actually getting applications in the first place.

It is not sufficient to increase remuneration rates, although thought has to be given to the relative competitiveness of the sector. Nor does the solution lie in a slick marketing exercise: leave that to the Armed Forces! The profile of this important area can be raised by a number of means. The following list, including many strategies already used, may stimulate thought:

- Road shows to schools, to present the various roles in housing to those in the last one or two years of school.
- Presence at recruitment fairs at universities and further/higher education colleges.
- Inclusion of housing management and strategy in cognate university degrees, and expansion of existing housing-related degree availability.
- On-the-job work experience schemes offering a taster of housing management to school leavers and job-seekers.
- Increasing the status of social housing through intelligent exposure in the media, through documentaries, soaps,

> broadsheet articles, chatshows, meaningful government initiatives, imaginative and customer-responsive housing schemes: anything really.

Think of the way that other public sector jobs have been advertised and promoted over the years. Nursing has been packaged as the 'caring profession', where the value of human concern offsets some of the financial disadvantages of choosing this career path. The armed forces are promoted as a means of learning a variety of trades and life skills, which helps take the mind off the imminent danger of violent death in a war zone. The police service is portrayed as an exciting mix of investigative work and human relations skills, and glamorised in innumerable soaps. Even the civil service has its own subtle means of promotion, through *Yes Minister*-type comedies and the antics of senior civil servants at various enquiries and state cover-ups; and there are adverts on the back of most London buses extolling the virtues of bus driving. So why not housing?

The marketing angle could be worked up around the central need to provide and manage housing for households who have been marginalised by the economic success of the country – the human interest angle. A career in housing should be portrayed as a way of meeting essential needs and contributing towards the economic growth of the nation by providing essential social infrastructure, with a focus on the variety involved in housing work – finance, strategy and planning, one-to-one dealing, investigation, legal work, customer and personnel relations and so forth.

An alternative angle is not to promote the field at all, but increase capital intensification by replacing all possible human operative functions with ICT solutions, following the automation of rent account and budgetary procedures. To downsize existing operations to provide the most basic of services, and to outsource as much as possible, on the assumption that it would be too difficult to construct a credible and workable recruitment strategy in the field, and next to impossible to retain personnel where there is so much choice elsewhere.

I am sure that this approach will never be taken.

References

Audit Commission (2004). *Housing – Improving Services Through Resident Involvement*. Belmont Press.

Blundell, B. and Murdock, A. (1997). *Managing in the Public Sector*. Institute of Management: Butterworth-Heinemann.

Chartered Institute of Housing (2001). *Repairs and Maintenance. Good Practice Briefing no. 22*. CIH.

Cooper, C. and Hawtin, M. (1998) *Resident Involvement and Community Action*. Chartered Institute of Housing.

Department of Environment/Welsh Office (1994). *DOE Circular 6/95. The Housing (Right to Manage) Regulations 1994*. HMSO.

DETR (2000). *A New Financial Framework for Local Authority Housing: Resource Accounting in the Housing Revenue Account*. HMSO.

Durkheim, E. (1936). *Suicide*. Routledge.

Green, H. (2004). *Staff Recruitment and Retention, a good practice guide*. CIH.

Housing Corporation (1998). *Black and Minority Ethnic Housing Policy*.

Housing Corporation (2003a). *Housing Corporation: Charter for Housing Association Applicants and Residents*.

Housing Corporation (2004a). *Involvement Policy for the Housing Association Sector*.

Housing Corporation (2004b). *Regulating a Diversified Sector*.

ODPM (2003a). *English House Conditions Survey 2001*. HMSO.

ODPM (2003b). *More than a Roof*. HMSO.

ODPM (2003c). *Options Appraisals, A Tenants' Guide*. ODPM.

ODPM (2004a). *Empowering communities, improving housing: Involving black and minority ethnic tenants and communities*. HMSO.

ODPM (2004b). *Supporting People: Review of the Development of the Policy and Costs of Housing-Related Support Since 1997*. HMSO.

ODPM/Cairncross, C., Morrell, C., Darke, J. and Brownhill, S. (2002). *Tenants Managing. An evaluation of Tenant Management Organisations in England*. HMSO.

Smith, I. (1997). *Meeting Customer Needs*. Institute of Management; Butterworth-Heinemann.

Bibliography

Blackaby, B. (2004). *Community Cohesion and Housing, A Good Practice Guide*. CIH.

Chahal, K. (2000). *Ethnic Diversity, Neighbourhoods and Housing.* Joseph Rowntree Foundation.
Chartered Institute of Housing (latest edition, updated twice-yearly). *The Housing Standards Management Manual.* CIH.
Chartered Institute of Housing (2000). *Managing Tenancies form Beginning to End. Good Practice Briefing no. 18*. CIH.
Chartered Institute of Housing (2000). *Best Value for Housing Staff: Good Practice Guide 17*. CIH.
Chartered Institute of Housing (2001). *Sustainable Lettings. Good Practice Briefing no. 20*. CIH.
Chartered Institute of Housing (2001). *Managing Rent Arrears. Good Practice Briefing no. 21*. CIH.
Cole, I., Iqbal, B., Slocombe, L. and Trott, T. (2001). *Social Engineering or Consumer Choice? Rethinking Housing Allocations*. CIH/JRF.
Coles, B., England, J. and Rugg, J. (1998). *Working with Young People on Estates: the role of housing professionals in multi-agency work*. CIH/JRF.
Duncan, P. and Thomas, S. (2001). *Neighbourhood Management: A good practice guide*. CIH/Housing Corporation.
HACAS Chapman Hendy (2003). *Empowering Communities – The Community Gateway Model*. CIH.
Housing Corporation (2002). *Communities in Control.*
Housing Corporation (2002). *Race Equality Code of Practice for Housing Associations.*
Housing Corporation (2002). *Making Consumers Count.*
Lupton, M. and Perry, J. (2004). *The Future of BME Housing Associations*. CIH.
National Federation of Housing (1998). *Equality in Housing.*
Nixon, J. and Hunter, C. (2002). *Tackling Anti-Social Behaviour: Action Frameworks*. CIH/JRF.
Robinson, P. (2003). *Leasehold Management, a Good Practice Guide.* CIH.

Websites

Essential websites for housing management issues for social landlords:

- Chartered Institute of Housing: www.cih.org
- Housing Corporation: www.housingcorp.gov.uk
- HouseMark: www.housemark.co.uk

- Office of the Deputy Prime Minister: www.odpm.gov.uk/housing
- Joseph Rowntree Foundation: www.jrf.org.uk
- National Housing Federation: www.housing.org.uk

Essential website for homelessness issues:

- Shelter: www.shelter.org.uk
- Crisis: www.crisis.org.uk

Useful website for information on resident participation, including good practice guides jointly prepared by ODPM and TPAS:

- www.tpas.org.uk

3 Housing finance

Introduction

Why do so many housing professionals decide to find housing finance uninteresting?

First, it is often equated with, or seen as a branch of, economics, which has been described as the dismal science, enough to put anybody off. Well, this is not really economics, although economics concepts will be mentioned from time to time to give depth and breadth to the arguments deployed. Arguably, economics is no more dismal than any other branch of study, and an understanding of the subject can actually assist in creating national wealth, and is certainly essential in the management of the country, often more revealed in the breach rather than in the action.

Second, many people have a very real fear of dealing with large sums of money, and express this in refusing to go into financial issues in any depth. Risk is everywhere, and wrong decisions can lead to the bankruptcy court or before the bench, or out of a job or a home. Many have had this unnerving experience, as well as the adrenalin rush of gambling and the sweet sense of success when a deal has gone well. If this is the case, do not worry, you are not being asked to make an investment, risky or otherwise.

Third, there is expertification. Finai
difficult, best left to professionals w
and who have taken finance or math
City or as accountants. Accountants a
they are a profession: don't try to m
which you are not qualified to unders
true that there are quite a few odd t
financial management studies, at roc
quite straightforward. Anyone who l
who has to pay bills, who has bough
drawn a wage or a benefit, or who h
withdrawal, can grasp the basis of fi
Like anything else, finance is foundatic
bottom up rather than trying to hang
and you'll stand a chance of understandi
fits together.

Fourth, it is dull. It's about bean countir
and far duller than the things and servic
isn't. *Rogue Trader* wasn't dull. Jeffrey Arc
Penny More, Not a Penny Less' is cracking, a
Money, Money' is arguably one of the l
twentieth century.

Fifth, it isn't about reality. Reality is the me
need, homelessness, bricks and mortar, c
old tangible nasty and nice experiences
can touch. It's nothing about value: every
accountants are people who know the pri
and the value of nothing. Not true. With
exchange, albeit abstract in one sense ar
objects of value in another, services would r
things would not get made, and this book
been written, because there would have bee
remuneration.

Finally, it isn't really anything specific to do
Again, denied. You can't build houses with
can build them for cash, and a lack of public
towards the sector helps to explain why the
unmet housing need even in the richer countr
Economic decisions have stunted the provision
ing, and have helped create the housing r

against spending commitments, as detaile
Spending Review. Tax is levied directly –
tax and corporation tax principally – and in
tax on commodities such as petrol, ciga
more or less everything else that we might
a good example of indirect taxation.

The government tries to act according to
and the Sustainable Investment Rule. The
spend only within current receipts (tax ma
borrow to finance current expenditure. In
within your means. The Sustainable Inves
keep borrowing at a level below 40% of
Product, which is the value of all product
tions in the country, less invisibles such as
The budget determines how much t
intends to raise from the public, and the
can spend is limited to capital reserves, in
ments, and tax receipts.

The government's policy on taxation is kno
as opposed to monetary policy, which is its a
controlling or influencing the money suppl
and monetary policy make up macro-econ

Broadly speaking, governments since the
monetary crisis have tried to keep public e
firm limits. The first reason is that they kno
a vote loser, and there is a natural tendenc
in power. Keeping taxes down and not usin
or relying overmuch on interest from inve
be done within the context of public e
Despite this, Public Sector Net Debt shows
ishing significantly, and budget deficits ha
underestimation of the need to spend on p
of the taxation required to do so.

Another significant reason for controllin
iture – through borrowing and cash-limiti
expenditure at central and local levels –
monetarist view of economics which ha
to a greater or lesser degree in the UK sin
This is the view that money is just like
modity, at least viewed from the stanc

economics. The more of it that there is around in circulation, if demand remains roughly level, or at least lower than supply, the less that currency is worth. If the money supply increases beyond demand, and is not backed by increased productivity, then it will be worth progressively less. This is inflation.

High levels of currency inflation are a bad thing, and tend to destabilise the economy, affect the value of exports, and decrease international confidence in the economy. In the mid-1970s, inflation levels rose above 10%, hitting 15% in 1977, and had a bad effect on the UK's international trading position, with one effect being cheap imports damaging home producers. It was a major contributor to industrial strife, with unions reacting to high price inflation levels by demanding equally high wage settlements. The so-called Winter of Discontent in 1978/9, the last year of the Old Labour Callaghan administration, was the result of public sector unions striking to secure pay deals in line with the high inflation levels of the day, and was the single most significant reason for Labour's defeat and Margaret Thatcher's ascendancy on a ticket of public spending constraint.

Although views now vary on the classical monetarist position, the essence of the approach still pervades Treasury thinking. The Cash Limits system, whereby departments were given strict expenditure and/or borrowing limits regardless of the need to spend, is still substantially around, although it has been modified to a certain extent since the late 1990s. It is fair to say that the government tries to keep Total Managed Expenditure – the total amount of money that the government budgets to spend every year – under strict control. It does so by holding departments of state, such as the ODPM, to expenditure limits on capital expenditure and borrowing to finance investment, as well as on predictable current expenditure – limits known as Departmental Expenditure Limits (DEL), and by trying to control Annually Managed Expenditure (AME) – such as social security benefits – through back-to-work policies and similar measures. AME is very volatile, and has proved difficult to regulate, due to uncertainties in economic performance, especially due to economic problems in some peripheral regions.

Prior to 1978, when the Cash Limits system was introduced, public expenditure tended to be needs-driven – witness the massive public sector housing programmes of the 1950s and 1960s, and the level of investment in nationalised industries and on regional development under the Wilson administrations. This was partly for political reasons, and also due to the Keynsian economic view that public expenditure was an excellent way of providing a stimulus to the economy by pump-priming industry and helping to create productive jobs which would in turn yield taxation income. The role of the state was seen rather differently then: the nationalisation of entire industries such as the railways, car production and utilities was seen as a way of guaranteeing jobs and economic growth through centralised, co-ordinated planning rather than leaving economic fortunes to the vagaries of the marketplace. It can be seen in the context of emergence from the wartime command economy, rationing, the foundation of the modern welfare state, and a reaction against the privations of the Great Depression of the 1920s and problems of the 1930s.

This was blown away by the negative experience of the 1978 Winter of Discontent, declining UK competitiveness against the Tiger Economies of the Pacific Rim, Japan, and New Commonwealth countries, and also by the manifest inefficiencies and non-market responsiveness of some nationalised industries to market changes, despite some very good examples of highly competitive nationalised concerns such as Rolls-Royce and British Aerospace. Mrs Thatcher and her government's denationalisation policies could not have been varied through had there not been very serious problems with the whole approach to running the UK economy manifest in the 1960s and 1970s. Successive governments in the 1990s and 2000s have made no moves to return to the days of nationalised industries and needs-based expenditure founded on high taxation and an acceptance of high levels of national indebtedness. Governments these days are enablers rather than producers, and are committed to ensuring a healthy private sector through trying to keep corporation and personal taxation down, taking a far more hands-off approach to credit controls and interest rates, and downsizing even its own administrative functions, as well as those of local government.

The philosophy of public expenditure control is manifest through the two-year Comprehensive Spending Reviews, introduced in 1998, which set out departmental expenditure limits in the context of actual and forecast national economic performance and the government's view of investment and current needs, but with an emphasis on the former. Each department has a Public Service Agreement (PSA) which indicates performance levels against which it will be judged.

The UK and European governments also find themselves having to tread with extreme economic care in the context of EU economic policy, and have increasingly less latitude for relatively risky macro-economic management.

In summary, then, housing finance exists within the macro-economic constraints set out above. Where public finance cannot be directly granted, any investment deficit must be funded through private finance: and the move away from the high Housing Association Grant regime to a very restrictive Social Housing Grant tariff since 1988 can be seen as a direct result of the concern to reduce public expenditure by increasing private investment. Witness also the Options of Private Finance Initiatives, the growth of Large Scale Voluntary Transfer and Arm's Length Management Organisations which can in many cases borrow privately, and the financial freedoms to borrow enjoyed by high-performing councils, and, of course, the Right to Buy, as indicators of public sector cash restraint and rolling back the central and local state.

Housing balances

Public finance constraint in the sense set out above provides a number of problems for housing policy and provision. It is acknowledged by most that there is a need to keep social rents below market levels to ensure that they are affordable for people on lower incomes, given the continued existence of sectoral and geographical wage differentials, and economic dependency through joblessness, age, disability and other factors. The evidence of the severe consequences for economic growth and performance of regions as a result of relatively high rents and owner-occupation entry costs is plain to see, and justifies sub-market rental and ownership solutions through state or state agency intervention through some form of subsidy.

The problem, then, is how to maintain a low or affordable rent regime and to provide reasonable management and maintenance services without undue call on the public purse, in line with low taxation, and against the backdrop of owner-occupation as the preferred form of tenure which, since the abolition of mortgage interest tax relief at source, receives very low levels of public subsidy. This is a balancing act which has tried most governments and shows no sign of going away. The other key problem is making sure that enough affordable houses are built without putting an undue strain on the exchequer to fund them.

Integrative overview – capital and revenue finance

Capital and revenue finance are fundamentally interlinked. Capital finance is essentially to do with funding the creation and enhancement (improving or adding value to) of capital assets, sometimes defined as things which are planned to last for more than one year, and exemplified in houses, roads and environmental improvements. Revenue finance is about meeting running costs, that is, current expenditure. The two are obviously linked: houses need maintaining, and loans raised to build or buy an asset (capital finance) need to be paid for largely by income from rents and interest (revenue), and managed and maintained on a day-to-day basis. Housing departments need to be staffed and staff expect to be paid (revenue costs), and they can't do their job without equipment such as computers, desks and chairs (all capital items, although some may be leased and hence treated as revenue expenditure).

If you are a home owner with a mortgage, you have bought your house with the help of a loan secured on the property, and have to pay it off over a term of usually up to 30 years, capital and interest or interest-only backed by an investment vehicle. The house you have bought is a capital item. The mortgage loan is a capital liability. The stream of mortgage repayments is revenue expenditure, and any interest you might accrue from savings put away for a rainy day or home improvements is revenue income, as is any income from a lodger. Day-to-day repairs and planned maintenance is revenue expenditure, as is the regular council tax payment and meeting bills. If you sell your house, the amount you have left after discharging the mortgage and other debts secured on the house is known as equity, and is a capital sum: and if you put it in stocks and shares, it's a capital investment which should generate an investment income (revenue income).

The same principles are broadly true for housing finance. The distinction between the two has sometimes been blurred mainly by local authority finance people, to try to get more money out of central government to meet housing need, and manage it.

Differences between the sectors

It is helpful to talk about capital and revenue finance separately for housing associations and councils, because the rules are very different. The basic difference between the way that the two sectors is treated is straightforward. Councils receive permissions to spend, or to borrow, to undertake capital work, except for major repairs – the control regime changes – and have to raise loans to finance this work, although they can also use some of the money they make from selling council houses. In many cases, they receive subsidy from central government to run their housing services and carry out running repairs and maintenance. Housing associations, on the other hand, if registered with the Housing Corporation (these are known as registered social landlords – RSLS), can receive social housing grant to meet a proportion of development costs, and have to borrow the remainder; and most have to meet running expenses from rents. Much of the rent money paid to councils and associations is backed by housing benefit, which is a personal subsidy to tenants on lower incomes, which receives central government subsidy.

There have been changes in the rules for both sectors over the past twenty or so years. Councils used to be allocated credit approvals, basically limits on how much they could borrow for all purposes, some of which was used for housing capital works, which were reduced by a proportion of council house sales, land and other receipts, and by so-called 'credit arrangements' like long-term leasing which was treated as existing borrowing. These limits were set annually by the ODPM and its predecessors, before 1 April 2004. Now, they are set supported capital expenditure limits based on a judgement by ODPM of their credit-worthiness and ability to repay loans, as well as on the council's assessment of its need to spend based on its Capital Business Plan. They are cash limits nonetheless, and define how much a council can spend on capital activities. Part of this borrowing is supported by government subsidy, by a stream of central revenue support and the rest is down to the local authority to finance through locally generated revenue (ODPM, 2003a).The total cash limits pot is determined in relation to other priorities in the Spending Review, which gives expenditure limits for three years. A three-year budget settlement was extended to local authorities at the same time as the result of the 2004 Spending Review, which makes capital planning much easier for councils.

The Housing Revenue Account, the current account of council housing which pays for landlord activities and receives rents and subsidies, has changed over the years too. The current rules governing the operation of the HRA are set out in the ODPM's HRA manual (ODPM, 2003b). Years ago, before 1989, it received government subsidy ('Housing Subsidy') worked out by adding an estimate of the amount spent additional to a base amount (uplifted annually by inflation from 1980 expenditure levels (the base year). Estimates of local income were then deducted from this sum, and Housing Subsidy was suppose to cover any deficit between base amount plus local costs, and local income. A lot of councils fell out of subsidy in the 1980s due to increasing interest income from Right to Buy receipts, to the extent that may became debt-free and could support themselves.

Then, in 1990, subsidy rules were changed. This was partly to ensure some control over council rent levels, due to some concern that they were too much lower than market or in some cases housing association rents, and that therefore unrealistic pricing signals for services received were being given.

Following the 'New Financial Regime' brought in from April 1990 by the 1989 Local Government and Housing Act, HRA subsidy was assessed as the difference between notional income and notional expenditure, with the government calling the shots on the figures going into the calculation, and including housing benefit in with the reckoning, so that some councils had to meet their own rent rebate (council housing benefit) bills, even though it is supposed to be part of the national income support system.

In 2004, this was changed again, with rent rebates being prized out of the HRA and passed to the Department for Work and Pensions, who began to pay councils for rent rebate losses to their accounts. Rent restructuring, informing the 'guideline rents' which ODPM assume councils charge, has also in effect reduced local democratic decision making on rent levels, and has had the effect of reducing HRA income resources in many cases. Finally, the way in which management and maintenance costs are assessed for subsidy purposes – the so-called management and maintenance allowance – was changed for 2004/05. Many councils, especially those with a high proportion of flatted stock, would have lost out significantly on their share of resources were it not for pressure from local authorities and their associations for transitional protection, although the 'targets' – the raw amount coming out of the formula – were in many cases lower than the targets before reformulation.

In the housing association sector, there have been correspondingly large changes. Prior to 1989, the amount that housing associations could pay towards development loans (from the government) was limited to the difference between the 'fair rent', pre-set by rent officers on a formulaic basis taking account of room numbers, location and so forth, and the cost of management and maintenance. In most cases, the residual amounts could only support loans covering a very small percentage of the development costs: and Housing Association Grant (HAG) was set to cover the remaining development costs. Some schemes, especially those for supported needs, attracted 100% HAG from the Housing Corporation, which was then a division of the then Department of the Environment (now ODPM). In addition, many housing associations received subsidy towards their running costs. The whole object was to keep rents down to 'fair rent' levels.

Everything changed from 1989, when the *1988 Housing Act* took effect, and introduced fixed-rate HAG. Fair rents were abolished for new developments, and grants were set at a fixed percentage of scheme development cost estimates called Total Cost Indicators (TCIs) without reference to rents, as a way of limiting the amount of cash going into the sector whilst cranking up development output. The remaining development costs had to be met through a development loan from the private sector – banks, building societies, and so forth – and paid for through rents which also had to meet the full cost of management and maintenance, at least in 'general needs' schemes. Because rent levels now arose out of development costs and grant calculations, rather than the other way about, it was necessary to abandon the regulated 'fair rent' regime for new properties and move towards unregulated assured tenancies at generally higher rents. Some tenants living in properties developed in 1988 with fair rents found themselves living next door to tenants in very similar homes developed in 1989 with higher assured rents, purely down to the change in grant regime. In practice, associations have tried to minimise the relative differences over the years, but gaps still exist, and it is very difficult to try to explain to a tenant why almost identical houses built just a few years apart attract very different rents, where the same differentials generally do not exist between council homes in the same authority.

The final major points which needs making is about the HRA as a 'ring-fenced landlord account'. It sits in the General Fund, but is a sub-account. It is unlawful to subsidise the HRA by dipping into the rest of the General Fund. This is the council account which receives council tax and revenue support grant, and which pays for council services available to anyone who needs or wants it in the community regardless of tenure, such as education, social services, planning, leisure and recreation, rubbish collection, and so forth – and private sector housing benefit, which is not available to council tenants. On the other hand, under some circumstances, the HRA must pay money into the General Fund outside the ring-fence – for example, when it is in notional surplus, even if it isn't. This restriction came in from 1990 as a result of the *1989 Local Government and Housing Act*, and had a dramatic effect on council housing management and maintenance service levels.

These are the general broad-brush themes, and hopefully helps to give some context to the detail which follows.

Capital operations in the social housing sector

Local authority housing capital finance

Capital activities include the construction and improvement of properties, as well as major repairs to them, and the setting out of hard and soft environmental features such as hedges, trees and fences. There will be more about the specifics in the next chapter.

Since the late 1980s, councils have built very few houses and flats, leaving it to housing associations to do so. The main reason is that the state pays directly for a percentage of the cost of building a housing association property through Social Housing Grant (SHG), with no ongoing impact on rents. A local authority would have to raise the entire amount, or commit right to buy receipts to a building project, without similar up-front subsidy. True, the loan repayment element would be subsidised, but with development loans having terms of up to 60 years, there is always the risk that subsidy withdrawal due to changes in policy would load the cost onto rents, or result in trimming some other housing service budget. Additionally, the ability of councils to borrow, and to spend their own receipts, has been restricted since the 1980s to the extent that many cannot afford to embark on the projects they might like to to create new housing. Finally, the backlog of major repairs required, and the need to improve properties to meet twenty-first century standards, has meant that councils have in practice prioritised improvement and major renovation schemes above new build.

A further reason concerns the now-defunct Local Authority Social Housing Grant (LASHG) regime, Prior to 2003, it was possible for councils to make social housing grant available to housing associations from their own resources, which would, in the majority of cases, be back-funded by the Housing Corporation. In these cases, the councils in question would use up a corresponding amount of their spending permission – their Housing Investment Programme amount would be reduced – but since grants are always less than 100% of development costs, that authority would be able to have access to more than one property per grant payment, compared to the one it would have to allocate to if it built it itself. Where LASHG was involved, councils usually got 100% nomination rights at least of first lets, so the property was equivalent to a council house as far as the council was concerned. The system was abolished partly because even debt-free authorities got back-payment to meet the cost of making SHG – money which could, and was often, used on non-housing purposes. Instead of simply changing the rules on the repayment of SHG, the government heavy-handedly abolished the whole system and put a number of pipeline development schemes in jeopardy, avoiding the worst effects by hastily devised transitional protection measures.

Annual Capital Guideline and Credit Approval

Until April 2004, Local authorities bid for permission to incur capital expenditure every year on the basis of a Housing Investment Programme (HIP). The HIP formed part of the authority's capital bid, across all service heads. They were granted permission to incur capital expenditure in January or February preceding the spend year, with indicative amounts for the following two years, and the housing part was within the general 'allocation'. The word 'allocation' is in inverted commas, because it was not a grant, rather an allocation of permission to spend. The allocation took account of a proportion of capital receipts from the sale of land, housing and other assets, in the bank to the start of the spend-year.

These allocations were expressed in the form of an Annual Capital Guideline, roughly a cap on the amount that the authority could spend on capital activities. The most important components of this were the Basic Credits Approvals – for general capital spending determined by the council – and Supplementary Credit Approvals (SCAs). These were amounts relating to expenditure on projects favoured by government policy. Some years, SCAs were issued to build hostels for the homeless, and in other years, estate improvement schemes were a priority.

Credit approvals were issued on the basis of the need to spend indicated by the Generalised Needs Index (GNI) , an attempt to weigh the housing need of one council against another using a form of 'points' system. Once issued, councils could borrow against these approvals, and spend up to their ACG by making up the difference by spending receipts.

Capital receipts

Councils have been forced to sell council homes to sitting tenants since the *1980 Housing Act* brought in the Right to Buy (RTB). Tenants of more than two years' standing as secure tenants of any councils have the right to buy the home they live in, with certain exemptions (for example, sheltered housing). They can buy with a minimum 32% discount, and can get up to 74% discount on flat sales subject to a ceiling, although the discount was restricted in 2003 in high-demand areas. Over 2 million homes had been sold between 1980 and 2004 on this basis, although the trend is now slackening – the best has already been sold off.

When the policy began, councils were allowed to spend all of the money realised, on housing and non-housing activities, such as doing up swimming pools, renovating piers and buying buses. Over the years, successive governments placed restrictions on the amount which they could spend in the year realised, partly because they feared that excessive municipal spending would fan inflation by increasing the amount of money in circulation, and to force councils to repay debts.

Up to 1990, they allowed councils to spend a percentage of the receipt – 25% in the case of housing cash, and 50% for land – in the year realised, and then to spend a same percentage of the sum left over in the following year, and so on, until the entire receipt was used up. This deferral system was known as the cascade. Councils could use the receipt to pay off debt if they wished. If they did so, the amount spent in this way, which could have been carried forward for capital spending, was known as a 'notional capital receipt'. These 'receipts' did not exist, and could not be spent. They came into their own in 1990, when government tried to stop councils from entering into 'long' leases on properties (of up to 20 years) in order to attract long term HRA subsidy. Such leases were treated as 'credit arrangements' – equivalent to capital expenditure – and their value was deducted from their total spending permission. Councils got round this by backing these leases with notional capital receipts, which had built up from 1981 to 1989.

The *1989 Local Government and Housing Act* effectively abolished the cascade, and with it, prevented the creation of new notional capital receipts and use of old ones after 31 March 1990, in an attempt to iron out some of the irregularities which had arisen through clever exploitation of the system by borough finance personnel. From April 1990, councils were only allowed to spend 25% of receipts from council house sales (called Usable Capital Receipts) and 50% of land receipts – forever – and the remainder – the 'reserved' proportion (Reserved Capital Receipts), had to be held in council accounts unless used to repay general debt, either on an enforced or voluntary basis. The receipts earned interest, the housing proportion of which were credited to the HRA.

One of the consequences of the RTB was the geographical unevenness of the distribution of their receipts. Councils in areas where the housing was popular or particularly desirable, or where there was in any case a string demand for owner-occupation, fuelled by house price inflation and the expectation of a substantial profit on resale after the end of the discount repayment period, sold much of their family-sized stock. They frequently realised enough not only to repay their housing and other debts, but to subsidise their general fund and HRA, keeping rents down, or embarking on municipal building projects. Debt-free authorities were exempt from restrictions on the percentage of capital receipts they could spend. Other councils, in economically depressed areas, remained reliant on the GNI and discretion-generated Credit Approval system. Many commentators remarked that the RTB gave gratuitous benefits to councils with low housing needs to meet and with stock in good condition, and none whatever to those in areas of high housing need and with significant need to invest in their less popular and unsaleable housing. Some called for RTB receipts, or at least the restricted part, to be pooled, and redistributed to councils and housing associations on the basis of area need.

Nothing was done about this until 2004, when the 'reserved' proportions of post-April 2004 receipts in England and Wales were pooled, and redistributed under the new prudential borrowing rules and in the form of SHG. Many commented that this was too little, too late, as RTB had by then all but run out of steam, and the rules were not retrospective. Councils in 2004 and 2005 still got the benefit of receipts reserved in 2002 and 2003, as the interest from them could still be used, and transitional rules meant that it was assumed for HRA subsidy purposes that they had not been use to reduce debt. In straightforward terms, HRA subsidy for debt repayment was higher than it should have been for a couple of years, as a softener (see ODPM (2003a) for rules).

The prudential borrowing regime

The Prudential Borrowing Regime, brought in under *Part One* of the *Local Government Act 2003*, came into effect on 1 April 2004, and affects the scope for councils' capital investment, including housing.

Councils now have powers to borrow what they consider they can afford to, guided by the Chartered Institute of Public Finance (CIPFA) Prudential Code for Capital Finance. Under the Code, prudential borrowing be affordable, prudent and sustainable, and is unsupported by dedicated central revenue subsidy. On housing, councils must take a view on the level of prudential borrowing supportable from locally-generated HRA income – basically, rents, weighed against other commitments. Because the flexibility to set and increase rents is limited by Rent Restructuring, most capital investment will probably continue to be financed through subsidy-backed resources 'allocated' from the local authority share of the regional housing capital pot by Regional Housing Boards, and through the use of the part of capital receipts not returnable to the government via 'pooling'. Prudential borrowing will be used as a supplement to centrally-backed resources, although it may help councils achieve their Decent Homes targets.

Centrally-supported borrowing is known as Supported Capital Expenditure, and the charge for capital is assessed by using a measure called the Capital Financing Requirement (CFR). These measures replace the old Credit Approval system limits. The CFR is divided between the HRA and General Fund, as the HRA is responsible for servicing housing debt. There is no doubt that the new capital regime is more flexible than the old one, but it remains to be seen to what extent councils will use these powers.

The advent of the Major Repairs Allowance (MRA) in 2001 marked a significant resource change for councils. (DETR, 2000). Prior to this, they had to use HIP resources to carry out major repairs programmes: since 2001, they have been able to charge this to MRA resources, allocated through an allowance to the Housing Revenue Account. This expenditure has to be planned, and appears in HRA business plans. The essential concept is that MRA is there to enable councils to replace components such as baths, WCs and heating systems, when they reach the end of their useful life in a power to obtain high-quality homes for all levels of owner-occupation on a discounted or full ownership basis in all developments to help create sustainable communities. The management of the social housing developments could be contracted to councils, housing associations or housing companies.

Why not? The new towns developed after the Second World War were developed by the Commission for New Towns, a central government agency, with public money, and provide a model for the finance of these developments. Again, it is only ideology which prevents the serious consideration of this very cost-effective option.

Financially, this option makes very great sense. There are significant economies of scale to be had from such large-scale commissioning, especially if the agency were the largest customer, or had a near-monopoly in this respect. This tremendous advantage could be used to drive down construction costs, as long as quality were monitored properly, with advantages to the public purse. The potential gains in tax income are tremendous: a national labour workforce would contribute a significant amount to the Exchequer directly through income tax, and indirectly through taxation on the goods and services consumed. It would also help absorb the demand for employment from new arrivals from the EU accession countries, on a permanent and temporary basis. It would also stimulate the employment market in Britain's depressed regions, and perhaps in some cases help to turn some of them round. This may sound somewhat Keynsian, but what's in a name?

Additionally, the large increase in housing supply coming out of these programmes would inevitably decrease demand, and with it, the price of housing in the private sector. So everybody benefits, apart from the development companies who have for far too long lived on excess profits, and whose investors could always try the stock market, paintings, rare books and stamps as speculation alternatives. Housing issues are far too serious to be subject to excessive international speculation risk.

It may be objected that this smacks of state intervention on a grand scale, and would involve the massive commitment

planned manner. It is not there to improve properties as such, nor can it be used to do this. It was generally welcomed by councils as an additional subsidy stream, and takes the pressure off capital resources to a certain extent, although the same amount could have been allocated as year-on-year spending permission.

Scope for reform

There is plenty of scope for different approaches to council housing capital finance. The preference for housing associations as development agents made sense when councils were unable to access private finance, but it makes little sense now that 'good' councils are able to do just that, and high-performing councils can use ALMOs to borrow privately to improve council properties. It matters little who develops social housing, as long as somebody does, and the presumption that others can do it better than councils is sheer prejudice. Councils can develop decent housing, as proved by their record in this area since the late nineteenth century. They can buy in the skills to do this important work, and contract with the best of them.

There are many advantages in moving back to council development – they are planning authorities, and it may be thought that internal relationships can be more straightforward and efficient than the present system, where housing associations are external applicants. There appears to be no good reason why councils cannot deliver quality to time, and why there should be an organisational separation between estimating housing need and delivering social housing. Councils are democratically elected bodies, responsible to their communities through the ballot box to meet all varieties of social need, and have the interests of all sectors of their communities at heart when planning to meet housing requirements, unlike housing associations, who often have interests dictated by the niceties of their constitutions, which may favour one client group over another, may operate over several regions and have to juggle development needs in one area with another within limited ADP resources, and are subject to decisions made by the Housing Corporation, which, although responsible to Parliament, is certainly no democratic organisation, and is not in touch with local need in the same way as councils.

Recent developments have proved that councils can be excellent enablers of development, through Private Finance Initiatives (PFI). PFI is essentially a vehicle whereby councils can guarantee private sector loans of 30 years for developers to construct or improve housing they own, contracting with a consortium usually made up of a bank, building society or other financial institution, and a housing company or association, to deliver improvements or construction over the life of the loan. PFI loans are backed by government subsidy, as set out in the annual HRA Subsidy Determination for the following year.

The council is essentially the client, and monitors the operation. PFI has been around for many years, and has been responsible for non-housing construction projects such as the Dartford Crossing, hospital construction and school building and renovation. It came into the housing arena in 2001, and has been used to support capital and revenue projects. It has its critics – there are obvious delivery risks, as 30 years is a long time contractually, and partners may default over that time period, but the record has been relatively good, so much so that it is one of the 'options' that councils choose in order to meet decent homes targets by 2010. If it is such a good scheme, why not allow councils to contract with others to deliver social homes on a much larger scale? There is plenty of need after all, if the government's *Barker Review (2004)* is to be believed. The only modification required would be to allow councils to be the developers themselves, rather than relying on a third party. Surely this would remove some of the risks of the project – by reducing the number of players. For a useful reference point on PFI Schemes, see OPDM (2004a).

Another reform would be to allow councils to use all their capital receipts, whether housing-generated or not, to construct social housing. The money is there: there is no need to borrow it, or to find security for loans. Monetarist concerns about increasing the amount of money in circulation are largely discredited, but can be allayed by considering

that whoever did the development would have t
use the equivalent amount of money to do the bus
why not councils? This could be achieved within t
ing redistribution of pooled receipts mechanism im
2004, to ensure that councils with larger resources b
lower needs contribute towards meeting the housing
of cash-poor authorities.

In the author's view, it is only ideology, not sound econ
ics, which prevents the return of councils to the deve
ment process. The scale of housing needs demands a rad
solution, beyond planning-led initiatives we have seen, a
councils, with their land reserves and historic experience
mass housing development, and lack of profit motive o
need to generate a surplus or to consider narrow sectional interests, are best placed to do this, especially given their transformation into quasi-businesses since the 1980s. What was CCT and Best Value all about, if not to make them business-ready: businesses with a social conscience?

Other approaches to housing capital finance which would enable much-needed housing development on the scale needed might include the creation of a national housing development agency, established on a regional basis. This agency would take over responsibility for house construction housing associations. Responsible directly to parliament, it would work with councils to establish and forecast regional and sub-regional unmet housing need, acquire land with wide ranging powers perhaps modelled on the *Community Land Acts* of the mid-1970s, use its buying power to drive down the costs of construction contracts, or even establish a national construction labour force to undertake the developments. It would have reserve planning powers to assemble land sites and ensure that NIMBYesque considerations did not get in the way of meeting housing need. It would, hopefully, incorporate a community consultative mechanism, working with councils to test local opinion on the scale and appearance of developments, and have a statutory duty to work with other departments of state and infrastructure providers to ensure that schools, hospitals, roads, leisure and recreation facilities were appropriately planned in. There is no reason why such developments should be limited solely to housing for rent: the agency could use its superior buying

of state finance with the risk that the return levels in the short to medium term may not match the outlay. To answer this it is enough to point to the state education and health system, and even to the departments of national government itself. These are hardly small operations, and are still essentially in the public domain. Nobody is seriously advocating the wholesale privatisation of these national resources, and considerable money and effort has been spent on trying to make them better. They are part of Britain's mixed economy. Why not housing, which is as much a basic social need as health and education. Is it not worth spending state resources on something which will benefit everybody directly or indirectly?

Arguments that the risk of insufficient return outweighs the costs of such an enterprise are questionable. People pay rent for homes, or purchase them using loaned or saved money, or a combination of both. A combination of fiscal policy re-alignment and re-apportionment of tax receipts at the margins and charging realistic rents to cover that proportion of development and management costs not met by the state, combined with selling the homes developed for owner-occupation at the going rate, could at least be modelled.

The level of unmet housing need, measured in the length of social housing registers; the unacceptable level of overcrowding in our major cities; households priced out of the market in urban and rural areas, and the obscenity of homelessness and expensive temporary accommodation use; these things stare us in the face. Their urgency is such that the onus is on the opponents of such possibilities to demonstrate how they would solve the housing need crisis, rather than for proponents of state investment on the scale required to defend themselves.

Housing association capital finance

Housing associations are independent, not-for-profit housing bodies. There are over 4,000 of them in England alone, of sizes varying from 100 000 to 10 homes. Of these, approximately 60% develop new housing. They have been around since the nineteenth century, and many have their roots in charitable or philanthropic institutions established by parishes building almshouses, or emerging from the troubled social consciences of great Victorian industrialists such as Peabody, Cadbury and Titus Oates. They are the only social housing developers – for rent and lower-cost homeownership – in the UK and the EU today.

The *1988 Housing Act* introduced the present capital finance under which they labour. As previously indicated, those which are registered with the Housing Corporation – known as Registered Social Landlords (RSLS) – receive Social Housing Grant at fixed percentages of Total Cost Indicators, and set rents accordingly, taking into account management and maintenance costs, and the need to build up a reserve fund for major repairs and improvements. The Grant regime has been modified to take account of limits on initial rent and increases implied by the Rent Restructuring regime, about which more later.

The basics of the grant regime are as follows. First, the Housing Corporation sets Total Cost Indicators (TCIs) which reflect the cost of developing properties of varying bedroom numbers and sizes, and dwelling types, which vary regionally. The Corporation then assesses the capitalised value of rents which would be necessary to support a given level of management and maintenance expenditure, and includes an element for a sinking fund to deal with major repairs and improvements over the life of the property type. It then deducts this figure from the typical TCI, and publishes the fixed grant accordingly, every year.

The development resource available is the Approved Development Programme (ADP) – the Corporation's capital budget voted on a three-year basis by Parliament, and emerging from the Spending Review. The ADP is divided up regionally, determined largely by the Corporation's assessment of housing need, which was (at 2004) guided by the ODPM's Housing Needs Index, and announced every February. At the time of writing, there were plans to scrap or reform HNI as a basis for division of ADP.

Housing Associations, under the 'traditional' route, bid for ADP resources, which is paid in three tranches (portions) – at contract signature stage, start on site, and completion, with some held back until expiry of the Defects Liability Period.

They then have to negotiate loans, or draw funds down from loan facilities already negotiated with the private finance sector, to meet the gap between SHG and total development costs. When doing so, they have to be mindful of the amount of rents chargeable and the degree to which these can be increased annually (dictated by rent restructuring), as well as management and maintenance costs and the need to pay into a sinking fund for future repairs. SHG is repayable only in the case of low cost home ownership, on sale of a 'share' of the value of the property to the purchaser. Previously, as already mentioned, local authorities could advance LA Social Housing Grant, but this has been abolished, although councils have limited powers to fund housing associations under their Local Government Act powers.

Many housing associations no longer receive SHG on a project-by-project basis, but a quantity of ADP resources to finance a whole development programme. Essentially they are given development targets in return for a guaranteed cash allocation. This is known as 'partnering'. This gives RSLs much more development flexibility than the grant-by-project regime, and erodes the utility of TCIs as other than overall indicators of programme value for money.

The development loans which housing associations have to raise are essentially fixed or variable rate products, and many associations use both. They obtain loan finance from a variety of commercial lenders, and can also raise funds on the stock market and through brokers such as the Housing Finance Corporation which is naturally sympathetic to the movement, but which nonetheless operates on a commercial basis. Lenders are now used to associations, and will lend both against the value of the asset, as is the case with domestic mortgages, or on the basis of income from rent.

Housing associations have to produce business plans, estimating income and expenditure profiles typically over 30 years, or however long the development loan period is for, which is sometimes considerably shorter, to convince lenders to come up with the loan. They have to convince lenders that they have firm but fair arrears policies in place, and are efficient managers, and can manage the 'risks' associated with financing loans. These include an assessment of likely arrears and (where applicable) their ability to service variable interest rate loans when interest rates increase (Housing Corporation, 2000, 2003).

Recently, their task in this respect has been made more onerous by the abolition of payments of housing benefit direct to landlords under the 'choice and responsibility' policies of the 2004 Blair administration, but they continue to attract private finance. They used to be able to, but can no longer, get loans from the state.

Size matters in obtaining loan finance. The larger the organisation, the better rate that can be obtained, because the risk of default is less, and the profit from lending is larger. Many associations have set up group structures, where several associations are constitutionally linked under an umbrella organisation, or have formed consortia, to raise money in a cost-effective manner. Housing associations are big business these days. Mergers are treated with great caution by both associations and their regulator, as governance responsibilities change radically with increased scale, bringing with them demands for board members able to understand and agree to complex financial arrangements, and assess the knock-on effects of diverse and more numerous stock (Housing Corporation, 2003).

Granted, associations are not-for-profit organisations, and are run by boards with a social conscience, even of they are non-elected. But, in playing the private borrowing game, they do contribute to the profits of private finance organisations, and if they weren't players, the private lending market would be that much smaller.

Scope for reform in the housing association capital area

Arguably, then, if housing associations weren't in the market for loans to the extent that they are, demand for money would be that much less, and if the number of suppliers or the amount of supply did not automatically adjust, the cost of borrowing would be that much cheaper, which would benefit business outside the finance arena, and domestic borrowers. There is always a lag between demand and supply changes, and finance institutions do not close down overnight or switch products instantly: the likelihood of reduced borrowing costs, even for a short while, would be significant. This argument could provide fuel for arguing for a return to the ability of housing associations to borrowing from the state. It would lend from current tax income and reserves against the very safe assets of property, or for a significant increase in grant rates so that the need for private borrowing, and filling the pockets of private investors, would diminish.

One solution which the above suggests is the ability of housing associations to borrow from the Bank of England, directly, or through a specially created intermediary. They would be allocated loan finance on the basis of a costed business plan detailing investment risk, income and expenditure over the life of a long-term 30-year loan: long term, because this would keep down the level of repayments. The Bank would take a charge over the revenue stream or capital assets of the association used to cover the loan, so that in default, the homes would become the property of the Bank, for transfer to another association more able to service the debt, or to a local authority able to do so through its rent and other income, or to a state housing body.

This would be advantageous for the state in several ways. First, it would guarantee that the development programme envisaged by Parliament actually takes place, without risk that lenders will not provide development finance, perhaps because more lucrative lending opportunities have arisen elsewhere. Second, the loans could be set at a level which would help guarantee affordable rents below current 'rent restructuring' levels, for public policy reasons, which might also create a saving in housing benefit (HB) depending on how the loan policy was pitched. Even if there were nil overall savings, there would be greater certainty in the Annually Managed Expenditure (AME) budget, as lower rents would certainly reduce the level of HB payments, and the Treasury would surely welcome the greater economic planning certainties implied. Third, the loan rates would be lower in any case, with favourable impacts on rents, because the Bank of England is a sizeable institution and could command loans from the private sector at reduced rates through volume of demand, and is not in any case reliant on private sector input. It also has the advantage of being in a position to set interest rates in the interests of macro-economic policy.

Another reform, already hinted at, is to increase the level of grant rates. One justification for this would be to hold rents down at affordable levels, and reduce the possibility of associations going under because of difficulties in maintaining income streams. Some would argue that the government has a moral duty to do this, as it may indirectly increased rent arrears through abolishing housing benefit direct payments in all but 'vulnerable' cases, if it extends the Local Housing Allowance policy to the social rented sector, which is expected sometime between 2005 and 2010. Another would be to reduce housing associations to exposure to the private lending sector, which cannot be guaranteed as a source of finance. This would entail a growth in the total ADP pot, with public expenditure implications, but if it resulted in a higher ADP output, the income generated in terms of taxation from associations, stimulation of the building sector with tax and employment implications, reductions in the Social Security portion of AME, may well out-balance the increase.

Another option would be to abolish SHG grant completely. Housing associations would develop at cost, subsidising costs only by built-up reserves or cross-subsidy from owner-occupation developments or market-rent schemes, and pass on the development loan, or residual amount, to rents, which would attract housing benefit. Instead of paying a capital grant to associations, the state would end up paying from AME. This may sound unlikely, but if real incomes continue to rise, with continually falling unemployment, and if it is believed that social security expenditure is set to fall, it may become feasible.

Revenue operations in the social housing sector

Council housing revenue operations

Council housing revenue operates under the rules governing the Housing Revenue Account (HRA), which is a ring-fenced account sitting within the General Fund (GF). The General Fund is the account which receives payments and pays out for services. The duty to keep a Housing Revenue Account has been in existence since the *1935 Housing Act*, but it has only been since 1989 that subsidy from the other parts of the GF to the HRA has been forbidden by law.

The HRA is a statutory landlord account, and contains income and expenditure items pertaining to the council's role as a housing provider. These items are prescribed by legislation, derived from the *1989 Local Government and Housing Act*, and modified over the years by regulation.

The main income items are rents – actual rental income (Item 1 credit), HRA subsidy, including management and maintenance allowance and the MRA (Item 3 credit), and interest from receipts (Item 8 credit). The main expenditure items are expenditure on management and maintenance (Item 1 debit), HRA contribution to interest payments on loans raised to finance housing by GF (Item 8(R) debit), and transfers to another account (Item 5 debit). Councils have a duty to endeavour to balance their HRAs, and can only carry deficits from the previous year forward to the next financial year. The HRA is subject to internal and external audit.

Changes in the recent past – from April 2004 – include the removal of rent rebates from the expenditure side of the account. Gross rents are now entered, and transfers from the General Fund of appropriate housing benefit subsidy are made to the HRA to meet losses from rent rebating, theoretically at 100% of rebates granted, plus a small amount to cover administration costs.

As stated previously, it is not possible to subsidise the HRA from the General Fund, except in respect of losses through rent rebates, and then only to the extent that the Department for Work and Pensions has made a payment to the GF. Prior to 1990, many councils did pay money from the GF to the HRA to keep rents down, and to ensure that services at levels decided locally were paid for.

Table 3.1 Entries to the HRA as prescribed by regulation

Expenditure (debit items)

Item 1	Management and maintenance
Item 2	Revenue contributions to capital
Item 3	Rents, rates, taxes and other charges
Item 4	Rent rebates (out from 1 April 2004)
Item 5	Negative subsidy transfer to General Fund
Item 6	Transfer to Housing Repairs Account
Item 7	Provision for bad or doubtful debts
Item 8 (R)	Cost of capital/impairment/deferred charges
Item 8 (S)	Depreciation
Item 8 (K)	Debt management expenses
Item 8 (F)	HRA set-aside (contribution to MRP)

Income (credit items)

Item 1	Gross rental income (From 1 April 2004, actual rents received)
Item 2	Charges for services and facilities
Item 3	HRA subsidy receivable (including MRA)
Item 4	Contributions towards expenditure
Item 5	Housing benefit transfers from General Fund (out from 1 April 2004)
Item 6	Transfer from Housing Repairs Account
Item 7	Reduced provision for bad or doubtful debts
Item 8	HRA Investment income/mortgage interest etc
Item 9	Transfers from general fund *(as directed by Secretary of State)*

(Technical items beyond the scope of this text)
Net cost of services
Debit Item 8 Adjusting transfer from AMRA (difference between cost of capital
and
impairment/deferred charges (R) and HRA interest costs (J))
Debit/credit Item 8 Amortised premiums (V) and discounts (W)

Net operating expenditure
Appropriations
Debit Item 8 (U) Transfer to Major Repairs Reserve *(where depreciation lower than MRA)*
Credit Item 8 (T) Transfer from Major Repairs Reserve *(where depreciation higher than MRA)*

(see ODPM (2003) for a full description and explanation of the debit and credit items)

Local authorities also have the power to keep a Housing Repairs Account, which sits within the HRA. The credit and debit items are Item 6 on each side of the account.

HRA Subsidy (Item 3 Credit, or Item 5 Debit)

Subsidy has been made available to help councils run their housing operations and help pay for loans raised to build housing since the early twentieth century. Amounts have varied, as has the principle of operation. The most recent system is based on a government view about what housing authorities should be spending to manage and maintain their stock, and on central assumptions about income, principally rents, rather than on the difference between actual income and expenditure.

There are good reasons not to use actual HRA figures to determine subsidy. Some councils are arguably not as efficient as they could be in letting properties, collecting rents, and controlling spending on management and maintenance, even though most have had over one hundred years' practice, and money from government applied as subsidy is, after all, a public resource which is not solely derived from the income of council tenants, even though, certainly in 2004, more was collected from HRAs by central government – effectively from tenants' rent payments – than was paid out.

Encouraging sensible budgeting and expenditure through subsidy limitation is one of several justifiable means of trying to ensure value for money in public services; but the way that subsidy has been calculated since 1990 has given rise to some criticism, and charges of unfairness.

Subsidy is calculated on the basis of 'notional' rather than actual HRA spending and receipts. The ODPM constructs an account for each authority based on its view of income under each head of the HRA, and its view on expenditure, and calculates HRA subsidy accordingly. This is a simplified account, for the purposes of exposition, and the reader who wants more detail should refer to the current HRA Subsidy and Accounting Manual, and HRA Subsidy Determination issued in December preceding each spend year.

The full specification of HRA subsidy is to be found in the Housing Revenue Account Subsidy Determination, issued each year. The formula for 2004/05 was:

> Amount of subsidy = (allowance for management + allowance for maintenance + allowance for major repairs + ALMO allowance + PFI allowance + Admissible allowance + Anti-Social Behaviour allowance + charges for capital + other items of reckonable expenditure) − (rent + interest on receipts)

Of these, the main items of notional income are rents, ('guideline' rents) and interest on receipts (i.e. capital receipts from the sale of houses and housing land), and main items of notional expenditure are expenditure on management and maintenance (the 'management and maintenance allowance' and MRA, and assumed interest charges for the HRA proportion of General Fund debt (crudely, 'charges for capital').

'Guideline' rents used to be assessed by multiplying a historically determined starting rent figure loosely based on 1989 rents by an inflation factor. They will, by 2012, reflect the rental income that the council would receive in the subsidy year if it had implemented the government's rent restructuring regime, with some adjustments to take account of the need to move rents towards the April 2011 target levels. The rents assumption is modified over a transitional period, in an attempt to ensure that subsidy losses or gains are gently factored in over a reasonable period, to avoid sudden resource losses or gains. The one thing that can be said is that these 'guideline rents' are not the actual rents charged. Rent restructuring will be discussed in greater detail later in this chapter.

Some authorities lost out on subsidy due to the introduction of rent restructuring, and a national compensation scheme called 'rebasing' was introduced to try to balance out the resource loss by adding an amount to management and maintenance allowances. Unfortunately, the balancing-out was on a national basis, to ensure that nationwide losses through rent restructuring were balanced out by increasing the national management and maintenance allowance 'pot'. Individual councils were not compensated for rent restructuring-related subsidy losses on a pound-for-pound basis, due to the way the distributional formula operated, and so in some cases, councils found themselves having to cope with actual HRA resource cuts. This was even though the national Management and Maintenance Allowances was uplifted by 6% from 2003/4 to 2004/5 and 2005/6 on top of smaller rises planned previously, on the basis of a government-commissioned study into the need to spend on maintaining council homes to meet a £1.5 billion shortfall in 2001/02 alone (ODPM/Building Research Establishment, 2003).

No allowance is made for arrears, although it is assumed that there was a rent loss due to 2% voids over the year.

Assumed interest from capital receipts is also entered on the credit side.

Management and maintenance expenditure is entered on the debit side, and assessed on the basis of a formula which takes account of the need to spend on responsive, planned and cyclical maintenance, varying with geographical area on the basis of building costs, type of property managed – there is a division into flats and houses and further subdivisions – and other factors such as wage levels and overhead cost variations. An attempt has been made to factor in the added costs of social deprivation, and the relative costs of communal area management. The formula to determine the level of assumed management and maintenance expenditure was last changed in 2003, and is good for 2004/05 onwards. The resultant amount is known as the 'management and maintenance allowance', and is entered into the debit side of the notional account. As with rents, there are transitional arrangements in place to try to ensure that resource losses or gains are factored in gradually. The raw management and maintenance allowance figures coming out of the formula are known as 'target' allowances. The worrying thing for many authorities, especially those with a high proportion of flats in their stock, is that the targets have moved downwards significantly since reformulation, and so much store is set on the continuation of transitional arrangements.

The total amount of M&M subsidy payable across England was calculated in relation to a study conducted by the government on the need to spend on M&M, in 2002. The division is on the basis of a distribution formula taking account of the factors outlined above. As mentioned previously in relation to rebasing, there was a recognition that there had been under-provision for M&M over the years, resulting in a large repairs backlog, and additional resources were made available under the terms of Spending Review 2002, but views vary as to the correctness of the distribution of the additional resources.

Interest on the HRA proportion of debt is also entered on the debit side, and assessed at the Consolidated Loan Fund rate, broadly the interest rate which it is assumed that the General Fund (GF) charges its constituent accounts which borrow from it, which varies between authorities. Only the council, as a corporate entity, can borrow money. It does so, and then lends internally to each service department at an internal rate of interest, which determines the rate of repayment. There have been some technical changes to the way in which loan debt repayment is assessed by the ODPM, relating partly to the change in treatment of capital receipts, but the technicalities are outside the scope of this book.

Other entries to the debit side of the notional account include the Major Repairs Allowance, which is the assumed cost of replacing building components to stock as they reach the end of their useful life, and MRA subsidy is justified by a Stock Conditions Survey, which the council has to carry out from time to time. It is based on the assumed replacement costs associated with properties, from 2004 divided into 13 types or 'archetypes', covering house and flat-types, and was introduced in 2001, as previously explained.

Authorities with Arm's Length Management Organisations (ALMOs) also receive an ALMO allowance, to help cover the additional contractual costs implied by contracting management out to another body. M&M subsidy is paid to the ALMO by the council. If they have PFI arrangements, they also receive a PFI allowance to back the cost of guaranteeing loans to the third party: and councils also receive an Anti Social Behaviour (ASB) allowance to help meet the costs of devising an ASB strategy, in line with current government concerns in this area. These latter allowance amounts are entered on the debit side.

The Admissible Allowance was granted from 2004/05 to partially compensate for subsidy losses relating to changes in the way in which capital receipts were dealt with, and is beyond the scope of this book.

Other 'reckonable' items of expenditure includes a variety of items, including (at the time of writing) payments for leases entered into before the 'New Financial Regime' imposed under the *1989 Local Government and Housing Act* of 1990 commenced, which were no longer covered from 1 April 1990. Their significance will diminish over the years, as leases entered into before that date expire, although some of them have a duration of twenty years.

Notional income is compared with notional expenditure, to work out subsidy entitlement. There are three possible outcomes: notional income exceeds notional expenditure; notional income equals notional expenditure; and notional income is lower than notional expenditure.

In the first case, it is assumed that the authority is in surplus on its HRA. Prior to April 2004, it had to pay an equivalent amount, known as 'negative HRA subsidy', to some other account, invariably the GF. The sum paid to the GF in this way was then generally clawed back by the Treasury through regulations applying to General Fund support. After April 2004, the notional surplus is paid straight to the government. Either way, the government gets some of the council's rental income!

This local-to-central transfer is known in some quarters as 'moonlight robbery'. It used to be called 'daylight robbery' before the subsidy rules changed in 2004, but amounts to the same thing. To the proponents of this view, the government is simply dipping its hand into locally generated resources on the basis of arbitrary assumptions about income and expenditure which have no basis in fact, as a convenient way of financing other objectives. It is viewed by these opponent of the subsidy regime as a tax on tenants' contributions, which could and should have been used to deliver housing management services, repairs and maintenance in the locality.

To adherents of the regime, it is simply a way of ensuring that the surpluses which councils' housing accounts would have made if they were operating prudently – that is, according to government guidelines on spending and income – are redistributed to finance central government objectives that will benefit more people than those who happen to live in the council's area where this ideal surplus should occur. As soon as a tenant pays rent to a public institution, which is what a council is, it becomes public money, and the government, legitimised by the sovereign parliament, is the ultimate custodian of public money. To supporters of this line, rental income is public money which may be locally generated, but it does not follow that it is therefore the property of the local authority. The debate can become legally quite complex, and will run and run.

In the second case, where there is a notional balance, the authority is assumed to be able to manage its own housing affairs without external help, and no HRA subsidy is payable, even if there is an actual deficit. Since rents cannot be increased to meet actual deficits, the only alternative is to spend less on delivering services. HRAs can therefore take a hit purely on the basis of the way that central government sees local housing expenditure needs.

In the third case, where notional income is less than notional expenditure, an equivalent amount of HRA subsidy is payable to the council. Due to the notional nature of the subsidy calculation, it may be more or less than actually required.

Since HRA subsidy is assessed on an annual basis, it is in practice very difficult to plan expenditure on services for more than a year ahead, which is hardly a basis on which to run a business-like enterprise.

Subsidy is paid in quarterly instalments, on estimates.

It is unlikely that the present system will be reformed significantly in coming years. It has settled down, and is immensely complicated, which in itself acts as a brake to anything but very minor change. It is very different from the pre-1990 system, which operated on a deficit subsidy basis.

Essentially, this took the previous year's audited expenditure and income figures, and added an assumed change in locally-generated income – mostly an increase, implied by interest from RTB receipts – and worked out subsidy accordingly. By 1989, the last year of the old system, most councils were out of subsidy due to increased income through RTB sales, and this was a source of concern to central government since it no longer had any way of influencing rent levels in those councils. Central government had tried, and failed, to do so through the *1972 Housing Finance Act*, which tried to impose 'fair rent' levels on council rents to bring some uniformity into the system, which would have had the effect of substantially increasing most council rents. Since the failure of this enterprise, subsidy limitation was the only way of controlling rents, and this was no longer available given the rise in local income.

Scope for HRA subsidy reform

Payment of subsidy on notionals would be more welcome, and certainly easier to justify to councils, if the notional figures which go into the formula bore some relation to actual or even reasonable expenditure and income – patterns of councils. Marginal adjustments of actual income and expenditure to calculate subsidy, in order to encourage efficiency to guard the public purse does not seem unreasonable.

In some respects, some of the items now bear a closer resemblance to the actual figures than they did, but only because the government has stated what the income levels should be in the case of rents, by imposing rent restructuring. More accurately, rent restructuring has not been imposed, but suggested. Councils which refuse to implement it are liable to lose subsidy heavily, so it's a form of Hobson's choice.

One option for reform would be to pay HRA subsidy on deficits for the coming year based on local authority figures. This would entail paying on the basis of HRA business plan estimates, perhaps quarterly. The end of the financial year would see councils auditing their HRA figures, with a check by an external body, perhaps the Audit Commission or Housing Inspectorate, checking figures against a peer group containing 'best practice' councils, perhaps those with 'Beacon Council' status under Best Value. An alternative would be a check to see whether the council had spent in accordance with prudent rules it had laid down and had accepted by the external body prior to the spend year. If councils could demonstrate that they have spent subsidy money and their own income prudently, no penalty would be levied on them. If, however, they had departed from their rules, or spent unwisely, the ODPM would have the right to claw subsidy back in the following year, or in other cases collect it from HRA reserves, or as a charge, on demand, within a month or so of the end of the spend year. That way, the realities of council housing expenditure needs would be recognised, with sanctions against poor performance.

The argument against this approach is that the government would never know how much it would be liable to pay councils on a year to year basis, whereas at present, the size of the subsidy pot is determined before the spend year, and divided accordingly. On the other hand, this is true of AME items such as social security in general and housing benefit in particular: why not treat HRA subsidy in the same way? After a few years of this regime, expenditure patterns would settle down, and the year-end checks and balances would encourage prudence and responsibility. There would still be councils whose incomes actually exceeded expenditure, in which case they would receive no subsidy; and there could still be some clawback and redistribution of actual surpluses if desired. At least councils could have greater certainty that they would be able to resource HRA business plans properly, and meet decent homes standards without having to be forced to consider farming out management of stock to other bodies, or disposing of stock altogether.

Another option would be to abandon the ring-fence on the HRA, and allow councils to subsidise their HRAs through payments from the General Fund, as they did before 1990. The key argument for ring fencing is that the HRA should be a landlord account which would stand on its own two feet: but most never have. All councils have a percentage of tenants on housing benefit – rent rebates. This is a form of national income support, and since 2004, subsidy to meet the losses incurred by rent rebating has been paid into the General Fund to cover notionally 100% of the loss. This sum has in practice been transferred to HRAs, to ensure that income otherwise lost can be used to manage and maintain council homes. Thus HRAs are already in receipt of subsidy coming into the General Fund. Prior to 2004, it was part of HRA subsidy. The HB subsidy element of the Revenue Support Grant settlement as now, or HRA subsidy pre-2004, is a form of general subsidy to the HRA. There is no contradiction between the HRA being a landlord account and receiving HB subsidy, which is an income support measure, just as landlords in the private or RSL rented sectors are no less landlords because the rents paid are often backed by housing benefit, which is external subsidy, and does not mean that they do not run landlord accounts.

It could be argued that the big difference between voluntary cross subsidisation from the GF to HRA and transfers of HB subsidy is that the former is locally generated and the latter isn't. This is patently untrue: first, the GF receives Revenue Support Grant (RSG) from ODPM to help councils ruin services to the general community, which is all general taxpayers' money, because there is no clawback and redistribution mechanism for RSG. Contributions from the GF to the HRA would in many cases be made up partially of RSG receipts which are not locally generated. Second, councils in notional surplus are forced to pay part of that surplus into the General Fund or to DWP into a fund which is recycled to help pay HB subsidy, and so part of the HB subsidy is in fact locally generated and redistributed anyway. So there is no good reason for prohibiting GF cross subsidy on grounds that it is sourced from local income from other than council tenants.

Another argument often used against abolishing ring-fencing is that there is no good reason why non-council tenants should subsidise council tenants, when they are not enjoying the services which tenants do. The argument runs that if councils did so, they would have to economise on general services which everyone can enjoy, which would hit non-council tenants disproportionately, because at least council tenants are enjoying another level of housing management and maintenance service than they would otherwise. Some have tried to measure the opportunity cost to non-tenants in the form of swimming pools not refurbished, the loss of amenity through parks not being maintained adequately, or the cost in terms of transport inconvenience when roads are not repaired.

However, this argument is unsustainable. There are costs to the community as a whole through inadequate housing management and maintenance often caused by lack of resourcing, which could be addressed through GF contributions. First, with the RTB entailing the pepperpotting of many estates – that is, creating mixed tenure estates of owner-occupied and council-rented properties – reduced ground and other communal maintenance forced by economies through lack of adequate HRA subsidy or the means to make good the deficit lead to lack of amenity to council tenants and non-tenants alike. Even where estates are not pepperpotted, lack of supervision and management could exacerbate unchecked anti-social behaviour on estates, which could spill over into surrounding areas, with implications for crime control and vandalism-related costs.

Lastly, although there are many more arguments which could be marshalled against retaining ring-fencing, council housing in an area should be regarded as a potential resource for all people living in the district. Anybody could require council housing at some point or other, through loss of income and ability to maintain mortgage payments, domestic relationship breakdown leaving them unable to buy or rent another home, and due to being a keyworker and not being able to access a pet scheme or being able to afford to rent or buy anything in the private sector. This local resource would fall to pieces if money was not spent on it, and would therefore not be available to the local community, or be available in very poor condition, and nobody wants this. Why should existing tenants be asked to foot the bill on their own for needs arising in the general community?

There is, then, no good reason why the GF should not be used in a limited and measured way as a source of support for the HRA. Many housing functions – the assessment of homelessness, provision of housing aid and advice, the warden service – are financed through the GF and backed by the Environmental, Protection and Community Services (EPCS) block of the Revenue Support Grant paid by government to support general fund expenditure on the basis that they benefit the whole community. So does council housing itself. So why not go that one step further and include HRA subsidy as part of EPCS?

Another option would be to abandon rent restructuring, and let councils raise rents to the extent that they need to meet the actual costs of providing council housing services. Councils are, after all, local democratic organisations. If local people, tenants or otherwise, wish to vote for a high-class housing management and maintenance service, perhaps on the basis of a menu of possible options, why not? After all, some councils have first-class swimming pools and recreational facilities because local people have voted for councillors who believe that these things should be supplied, and local council taxpayers are happy to share the cost of these superior facilities, and may even take some civic pride in them. It may be objected that this would mean that in some areas rents would be very high, or higher than they are now. This does not necessary follow, if councils are allowed to subsidise rents by GF payments, which would be reasonable if local people voted collectively for excellent housing services, which they might well do in areas of severe housing shortage through other routes, and where they or sons and daughters might have to rely on the sector for affordable housing depending on the degree of housing association activity in the area.

Even if rents were higher as a result of local decisions, there is no reason why those unable to afford the cost of the full rent should not continue to enjoy support from housing benefit, which could itself be reformed to endure that the withdrawal taper is not so steep as to make people better off if they continue to claim than if they take lower-paid work. It could be argued that higher rents would merely be a reflection of desire for a higher standard of service and repairs. The beauty of letting rents meet the cost of services is that it might even reduce central government subsidy levels, and would at least allow government to claim that it was encouraging tenants to pay for what they get, which is something which governments of all political persuasions have been trying to signal for years.

The bottom line is that ringfencing in the form that exists now only makes moral sense if it also applies to HRA contributions to the GF. It does not. Therefore ringfencing makes as little moral as it does financial sense.

A final option would be to use councils explicitly as government rent collection agencies, and for the centre to pay councils what it considers necessary to run housing services and do repairs and maintenance, and to finance loans raised to build houses and undertake capital works. This would, however, require primary legislation and fundamental changes in the constitutional status of councils. It would also fundamentally erode local democratic control and governance, although some would argue that things have already moved close to this position, given the amount of HRA money which flows back to central government, for centrally prioritised redistribution.

Rent restructuring

In 2002, the government decided to act on its 2000 Housing Green Paper *Quality and Choice in Social Housing* intentions to encourage councils to set rents on the basis of attributes of properties which tenants value, and to help ensure that all rents for similar properties in the social rented sector, regardless of ownership, were similar in level. It devised 'rent restructuring', a voluntary system for councils and RSLs, where rents would be worked out on the basis of value, property size and local income levels, and would move in a series of equal steps towards the target implied by the formula by March 2012. The system would be subject to a series of three-year reviews.

It was also proposed to base the 'guideline rents' part of the HRA subsidy system on the position which local authority rents had reached on their way to target rents, albeit modified by 'damping' assumptions to ensure a relatively smooth transition towards the targets. Finally, it was suggested that councils should consider 'de-pooling' the service charge element from rents, so that tenants living in blocks and estates with communal facilities not enjoyed by other tenants should have to pay something towards some of the unique features and services they enjoy. The 'rent' was therefore the tenants' contribution towards the costs of liabilities incurred by the HRA net of block or estate-specific services, such as repayment of HRA loan debt, the cost of general housing management and maintenance, and non block-specific staffing costs.

The government imposed strict limits on how much rents and service charges could rise by each year, as well as specifications for working out the income-related and valuation-related elements of the 'rent' element of what became known as the weekly payment. Councils did not have to adopt rent restructured rents, but it was made clear that it was preferred that they did, moving towards targets by 2011/12, and with additional incentives to de-pool by allowing councils to keep much of the income derived from such charges (see DETR, 2001).

For its part, the Housing Corporation encouraged housing associations to move towards restructured rents, and based their grant calculations on the assumption that they would do so. This, and especially the limitations on rent increase aspects of the policy, gave concern to private lenders who had in many cases negotiated loans on the basis of 30-year business plans devised long before the advent of rent restructuring. Large Scale Voluntary Transfer (LSVT) landlords in particular found their relationship with lenders becoming strained, as many had promised to impose 4% or higher annual rent rises on tenants, whereas rent restructuring limited such rises to RPI +/− £2 + ½% which was in many cases lower than the figure negotiated with financiers on transfer.

Despite some misgivings, most social landlords had implemented a form of rent restructuring and de-pooling of the service charge element by 2005. It proved difficult to ensure convergence of target rents implied by the formula and actual rents, due to limitations on increase levels, and the vastly different starting points across the council and housing association sector, and still more difficult to 'harmonise' rents – that is, to reach a point of genuine similarity between properties in similar areas of similar characteristics and value owned by RSLs and councils. The mechanism was poorly understood, and often mis-communicated to tenants. It did, at least, lead to more consistency in rent-setting policy than there had been before; but the assumption that guideline rents for HRA subsidy purposes would be moved up annually to reach restructured target levels by 2012 forced many unwilling councils to set rents in this way or lose out on subsidy, and therefore eroded a degree of local democratic control over the HRA.

The attempt to influence council rents is nothing new. Under the *1972 Housing Finance Act*, councils were told to set rents at fair rent levels, to minimise the difference between private sector and public rental levels. There was a storm of protest, and that part of the Act was never enforced. The 1980s saw ministers urging councils to set rents with differentials based on variations in the value of different property types in the area, and trying to reflect these differentials in the subsidy regime to force the issue. Then, from 1990, central government tried to force councils to set rents at levels compatible with central government aspirations by modifying guideline rents accordingly, but giving up in the mid-1990s and simply applying an inflation variable to base guideline rents. Rent restructuring is the latest in a long line of measures to impose some form of central control on local council rents.

The underlying reason for this is to control the level of that most volatile part of AME, housing benefit expenditure. If councils had the power to set rents just as they liked, assuming that HB was not limited in any way, there would be little or no control over the HB budget in respect of rent rebates. Central control over rent setting is one means of helping estimate the total annual HB subsidy requirement. It also means that this part of council income, and thus assessment of the ability of councils to fund their own housing management and maintenance on a year-on-year basis, can be more accurately measured. Crucially. It also gives central government more control over local housing expenditure, and thus a macro-economic control lever in its attempt to steer the economy.

Central government has in fact used other means to limit the amount of HB subsidy going to councils, by not paying anything over the so-called 'limit rent', which has converged with rent restructuring levels over the past few years, but which is logically independent of this. Limit rents effectively act as a barrier to councils charging significantly more than a given amount in the expectation that the increase will be met at least in part through HB subsidy or that part of the GF transfer attributable to HB backing.

It is an open question as to whether rent restructuring will in fact achieve any of its stated aims by 2011/12, but one thing is clear. If it doesn't, it is a fair bet that some other means of attempting to control the rent-setting policies of local councils and independent housing associations will be devised for all the reasons given above, in what has become something of a cat-and-mouse game of move and countermove, with the identity of the cat not always clear.

Which subsidy – capital grant or revenue?

Councils receive HRA subsidy to help them meet the costs of running a housing service and paying off loans, and no capital grant. Housing associations receive an up-front grant to meet part of development costs, and no subsidy to help pay for the residual loan, management and maintenance costs. Is there a best way of subsidising social housing?

It is undeniable that social housing must be subsidised in some way, in order to keep rents affordable. If social landlords had to meet the entire running costs and development loan charges associated with providing and maintaining housing, the rents would be unaffordable for many customers without significant housing benefit backing.

There are clearly options as to how to ensure that rents are affordable, and it is possible to approach the issue by examining the extreme solutions of all-capital or all revenue support, and examine the ongoing costs and benefits to see if there is a best way of doing so, or whether a mixed solution might be better.

No subsidy

Consider the situation where there is no subsidy at all. New housing developments would have to be financed by loans or reserves. The full costs of management and maintenance would accrue to the landlord. The ongoing revenue costs would have to be met through income generated on the schemes themselves. Some landlords, with older housing where the loans have already been paid off, and where economies of scale can be obtained to keep management and maintenance costs on those properties low, might be able to achieve comparatively low 'cost rents' to occupants, where 'cost rent' is literally the total revenue cost of providing the property. A judgement could be made that tenants could afford to pay more than the cost rent. The surplus generated from the rents over costs of the stock could be applied as a form of internal subsidy to the costs associated with new or newer properties, to bring their cost rents down to something like the level of rents on older properties.

Such a landlord might also be able to subsidise cost rents by selling some properties or other assets, and using the interest from investing it to offset some of the costs, thus reducing rents even further. It might even go in for development for sale, and invest sales profits in a similar way. It might even offer its management and maintenance services at commercial rates to other organisations or to individual customers, and use profits for rent subsidisation.

This form of cross subsidy is used by some housing associations, who apply surpluses from low-cost home-ownership sales to the development and running costs of for-rent schemes to keep rents down. Some have boasted that by using internal cross-subsidy, they can do without development grant at all. It would be possible to undertake a study to see whether and how many social organisations could keep rents at affordable levels without need for external financial assistance. The fact that grant is still available to housing associations in these hard-nosed times indicates that someone has done the calculation and found that this would work only in a minority of cases. And management and maintenance costs on older stock are often significantly higher than on newer properties, so the cost rent savings on older properties with expired development loans might actually be lower than required to make much of a dent in the rents of newer homes. It's still worth costing out.

Income subsidy only

Instead of subsidising bricks and mortar, and running costs, it is possible to subsidise customers who would not otherwise be able to afford the cost rent. In practice, this would mean a considerable increase in the housing benefit bill, but the size of the bill would depend largely on prevailing economic conditions. If real wages increase due to an economic upturn, the housing benefit bill would reduce, as more people paid more of the rent themselves. In times of economic decline, with rising unemployment, the reverse would be the case, but in the long term, it could be seen as a case of macro-economic swings and roundabouts. This could be seen as a disadvantage: AME is very volatile, and would swing around considerably due to economic booms and slumps, and collapses and growths in regional and sub-regional economies. All governments prefer to maintain a high proportion of planned over all expenditure, if only for fiscal reasons. For such a policy to work, there would have to be considerable reform of the HB system, especially in respect of the withdrawal rates, to ensure that people are not trapped in benefit-dependency, unable to take lower-paid jobs due to losing more income than gained, but reform of HB is nothing new. Again, it is possible to cost out this option. One modification of this approach would be to encourage landlords to set their rents per property at or near the cost rent level of a new property of that size, which would include a substantial contribution to loan debt repayment, across their entire stock. Landlords with a comparatively old stock, with over half with no further loan debt liability, would make a surplus on that portion of their stock. The government could then sequester this surplus in the form of negative subsidy, and recycle it in the form of housing benefit. Rents would then be high for those who could afford it, otherwise not.

Capital subsidy only

Another approach could be to subsidise the entire capital development and major repair and improvement costs by way of a grant, to all social housing organisations, whether council or housing association. The revenue costs – of management, responsive, planned and cyclical maintenance costs – would then be met through rents, which could attract personal subsidy through HB in cases of personal unaffordability. The advantage to central government would be that total capital expenditure could be planned at the beginning of each financial year with some certainty, with some contingency for unanticipated major repairs liabilities, although organisations could be required to generate a sinking fund against such eventualities. AME would be no more volatile than at present, and may even be lower, since associations, receiving 100% capital costs subsidy, would be able to charge lower rents than they do at present on SHG-supported development. There would be ongoing loan servicing costs for central government, where the cost of capital could not be met out of reserves, but these costs would be largely offset by savings in housing benefit and revenue subsidy. Again, this option could be costed out, and may not contradict the aims and objectives of the Prudential Borrowing regime's 'freedoms and flexibilities'. Letting councils use a grant to support capital or revenue operations would be a sign of central trust in high-performing councils.

It is necessary to take a fresh look at subsidy, to examine the overall effect of mixed and pure approaches to see what the best case would be for tenants, for social housing organisations, and for central government, without imposing cathartic change on the sector. The present subsidy systems are not the only ones available, and have arisen through historical reasons rather than because they are ideal or pre-ordained, although incremental change is probably the best way of implementing reform so as not to impose undue strains on an already shaky sector.

Housing support

At the time of writing, around 60% of all local authority tenants were in receipt of housing benefit, and the housing associations exhibited similar dependency. It is not surprising, because social housing landlords accommodate mostly poor households. If they were not poor, then they would probably become owner-occupiers, or so the orthodoxy goes, due to the relative advantages of that tenure.

As well as providing a roof, owner-occupation has been from time to time a good medium-term investment option for around 70% of the UK population. It is also a means of borrowing cheaply against owned equity as mortgages reduce as a proportion of value due to house price inflation, and the security is very durable and relatively immovable bricks and mortar. It is a more fashionable tenure than renting, the preferred form of tenure, as the enthusiastic take-up of the Right to Buy in the 1980s and 1990s demonstrated.

Housing benefit is a form of social security, a part of the national income support system, and available only to those who pass a 'means test' on income and capital. It increases with rent level and lack of income, and decreases as rents fall and/or income levels rise. It has been subject to reform in recent years; whereas in the past it was set on the basis of actual rents, or levels of rent determined by the rent officer service before modified in relation to means, it is now set on the basis of average rent levels for that type of property on an area basis. At the time of writing, Local Housing Allowances, set in the latter way, are replacing the rent determination model in the private rented sector, and set to replace the property-specific HB system which applies at present in the social rented arena (see Zebedee and Ward, 2004).

When should housing benefit kick in?

Eligibility for housing benefit depends on meeting certain income and savings criteria. If the claimant's income is at or below the income support level, then they are entitled to full housing benefit – that is, benefit at the top of the applicable scale, given the nature of their dependent household. If their savings (capital) are above a given threshold, which is varied from time to time, then they will not be entitled to the benefit. Capital is treated as if it were invested, and the notional income produced is deducted from the amount that can be claimed to a limit.

Income support levels are determined by considering what a claimant, given the nature of his or her household, needs to live on. This amount is varied annually, in consideration of the increase in prices of necessaries such as food, bills and clothing, and feeds directly into the housing benefit systems, although there are subtle variations in determining levels for this purpose. The assessment of the so-called 'applicable amount' – the amount which a household is deemed to require to live – is in principle difficult. How is it possible to decide what households need to buy and which services must be used to maintain an acceptable lifestyle, and acceptable to who?

It might be possible to start with a consideration of the basics which all households need. These include food, clothing, a secure place to live, medical care when needed, enough money to use at least public transport, cleaning materials, and so forth. How much of all of these do households consume? Surveys are carried out from time to time to determine what people actually spend on what, and at least some of the results are used by government when reviewing income support levels. It sounds relatively straightforward: but take food. How much does it cost to buy enough nutritious food to feed a household, assuming that the food is adequately prepared, which presupposes a level of skill and access to adequate cooking equipment? Is that food readily available locally? Dietary requirements vary: how does the income support system take account of this? Then there is clothing. Theoretically, one could fill the wardrobe with clothes from charity shops, and some are in good condition. This would be a lot cheaper than buying new, and the quality of the second-hand garments may in many cases be better. Should households in receipt of income support benefits be asked to use charity shops, and the income support element towards clothing be based on the price of a basket of clothes from Oxfam? If not, and I am certain that the idea of forcing people who are unable to provide for their needs from their own income to go to charity shops would be regarded as unacceptable by many if not most people, no arrears can arise through the rebated part of the rent, although administrative problems have entailed significant 'paper' arrears, where HB departments have failed to liaise effectively with rent accounts sections in notifying changes of housing benefit eligibility and amounts. Such mistakes have regularly been punished by subsidy limitations and on re-imbursement of 'overpayments' by councils arising from non-notification of changes of circumstance. There has also been benefit fraud, where claimants have failed to notify changes of circumstances or made bogus claims – but this happens throughout the social security system, and is not a special feature of rent rebates in particular, or housing benefit in general. If the system worked perfectly, all losses through rent rebating would be re-imbursed to councils.

Prior to April 2004, notional surpluses on the housing subsidy element of HRA subsidy, which included subsidy to meet losses through rent rebating, were clawed back through reductions in the HB subsidy element. In some cases, councils with rent rebate losses received no HRA subsidy at all, and in these cases, it could be said that rent rebates – part of the national social security system – were paid for through other tenants' incomes through rent rises, or cuts in services. In the first case, tenants were being asked to pay for services enjoyed by those unable to pay. In the second, those paying rents were being asked to accept potentially lower service levels for the same rent level. Thus, in these cases, rent rebates were a form of local poor relief of the form introduced by the eighteenth century Poor Laws, where destitute people were the responsibility of the parishes in which they lived. The situation was even worse for councils which had an actual HRA deficit, or which broke even, and were assumed to have a notional HRA surplus. In these cases, the legal requirement to rebate entailed cuts in services, or general rent rises falling to the pockets of those not eligible for HB.

The present subsidy system is rather more benign. HB subsidy is now taken out of the HRA, and losses to rent rebating are no longer accounted as a debit item. Instead, actual rents received are accounted as a credit, and the difference between this and the full amounts which would have been received if no rebating had taken place are credited as a

including the claimants themselves, then what quality of clothing should households be 'allowed' to consume?

Cheap clothes are often a false economy, and fall apart more rapidly than those which are more expensive initially: but should society be asked to pay towards clothing for those in receipt of state benefits which those on low wages might not be able to afford, on the basis of replacement cost and interval?

If yes, then economies could be made, because if the replacement interval is increased, say, twice, then the clothing element in the income support calculation could be reduced by a factor of, say, 25%, on the basis that clothing which lasts twice as long as the cheapest clothing (assuming that this is the present basis) will cost 1.5 times as much.

Against this, it could be argued that by 'forcing' claimants to buy more expensive clothes on the basis that the replacement interval will be longer, even if you could do this, would inflate clothes prices, as sellers react to the increase in bidding price for their clothes, which would in turn tend to increase the clothes element of income support (IS).

The assessment of the income support level is made more complex by the variation in the nature of households. It should be said that the sort of calculation made above is not actually representative of any official reckoning, and IS levels are pretty arbitrary. How do you calculate the income support requirement for a household with children and a

Table 3.2 Table of necessaries for HB

Calculation of reformed clothing allowance:	
(a) Cost of cheapest clothing:	£10 = pre-reform clothing allowance
(b) Length clothing lasts:	1 year
(c) Cost of equivalent clothing lasting 2 years:	£15
(d) One-year cost of more expensive clothing:	£15(Price of article) /2 (last-length)
(e) Reformed clothing allowance therefore:	£7.50

pet? Why shouldn't households on income support have pets? They are said to be very therapeutic, and many on IS may be unable to work due to disability, and may find the presence of a cat or dog good company and very sustaining. Supposing one of the children is very intelligent, and would probably be a major contributor to medical science if he or she had access to a computer at home, in case bright ideas emerge then and there, and require fuelling by an Internet search? It is undeniable that society needs bright scientists to advance medical research to benefit all. The risk of not providing a computer for such a child might be counted in lives lost through failure to nurture the child's particular and perhaps purely personal brilliance, through failure to give adequate collateral support when needed. Where is the line to support to be drawn? Can it, or should it, be delineated?

Just before we leave this debate, it is too easy to say that clearly the line must be drawn in order that AME can be managed within the context of macro-economic steerage, in the interests of all, and that the country cannot afford to pay for every eventuality, nor should taxpayers be expected to subsidise non-taxpayers to the extent that their lives are fuller and more productive than their own. Further, it can (and has) been argued that comparatively high levels of income support discourage work, and devalue the effort of the lower paid, to the detriment of all, and that 'choice and responsibility' means helping people back into work so they can make a contribution to their and everyone else's welfare. The odd argument is sometimes heard that, if everybody was in receipt of social security, then nobody could be, since there would be no income coming into the state to pay support, and the slightly less ridiculous argument which attempts to calculate a critical total maximum income support pot by assessing the amount which could justifiably be raised from taxpayers, weighed against other demands; for example, the defence bill, and the salaries of bureaucrats charged with working out social security amounts.

Hopefully the above shows that the issue of assessing minimum levels of household expenditure required to live an acceptable life, even without going into the issues around

the specific requirements associated with different ne characteristics, is not straightforward. At the end of the d the government has a democratic mandate to set incor support thresholds, and the reader is referred to the DW website, or to one of the many excellent annual referenc books on housing benefit rules and regulations – see the lis of useful reference sources at the end of this chapter.

Types of housing benefit

Rent rebates

At the time of writing (April 2004), council tenants eligible for housing benefit receive it in the form of a deduction in the rent they are required to pay. The deduction is based on standard housing benefit rules. Prior to 1983, there were different sets of rules for this benefit, depending on whether the claimant was a council tenant or private sector tenant, and on whether he or she was in receipt of income support or not. The rules were fully unified as a result of the *1986 Social Security Act*, as amended. The more they earn over the applicable amount, the less the amount of the deduction. In 2004, the standard withdrawal rate was 65p per £1 earned above the applicable amount. Benefit entitlement is reviewed periodically. Similar rules apply to Council Tax benefit, although the actual support levels and withdrawal rates differ. The loss to the HRA is met at roughly 100% of the amount by the government, and comes from the Department for Work and Pensions budget. The subsidy is paid to the General Fund, and made over by the council to the HRA as a transfer.

At the time of writing, there were moves to reform this form of housing benefit. It should be said that the term 'rent rebate' is no longer used for council housing benefit, which is known as housing benefit (public sector) except in housing finance circles, or just housing benefit. In 2004, housing benefit was deducted as a rent rebate on the basis of actual rents, once the means test had been applied. Rent rebates represent a guaranteed income to housing accounts. You cannot not pay a benefit which is never paid out in the first place; there is no choice but to pay it. Therefore, in theory,

transfer from the General Fund, which receives subsidy from DWP. However, the notional HRA still exists, and if a council is judged to have a surplus on its notional HRA, it has to transfer an equivalent amount to the government. Thus, even though 100% of rent rebate losses are credited to the GF by central government, and the full amount is paid over to the HRA, government can still claw back some of that HB subsidy income in cases of notional surplus, even if the HRA is just balancing, or in deficit. In these cases, it is apparent that councils are losing some of the HB income they need to run services due to government clawback – and councils are unable to make up the difference by increasing rents, due to the increased limits imposed under the rent restructuring regime. It is possible for some councils to lose roughly as much as they receive from central government to their GF to meet rent rebating losses. In these cases, it could be argued that, since councils still have to reduce rents, the reductions are being financed by cuts in service levels or actual services to all tenants, and so all tenants are being robbed of service levels which they would otherwise have received.

Remember that it is not possible for councils to decide unilaterally to increase the amount transferred from the GF just because of the difference between notional and actual HRA amounts entailing a loss of HRA resources. This would be a breach of the 'ring-fencing' rules already referred to. This is an ongoing problem for councils to wrestle with.

Governments have argued from time to time that there is really no unfairness here. Someone has to pay for social security provision. In these cases, the bill is partially met by the citizen at borough level, rather than by the general taxpaying populace. The problem with this argument is that this system imposes a larger burden on local people than it would if the costs were distributed over the UK taxpaying population, and fails to answer the objection that all other forms of social security are financed nationally rather than locally, and that there is nothing special about housing benefit which naturally means that it should be financed by local people in an antiquated and long abandoned 'poor law' fashion.

Imagine the situation which would obtain if all social security payments were to be treated in this manner. An assessment of government-calculated local income via a portion of locally generated income tax revenue and council tax revenue could be made, on a formula basis, to give notional income. Against this could be set the government-assessed estimate of what it should cost to run not only council services, but also all locally administered services and payments, such as income support and health services. If notional income exceeded notional expenditure, no central support would be payable to support security payments, or the excess would be clawed back from the 'area resources grant'. Either way, local people would end up paying the social security requirements of other local people. The government could argue that someone has to pay these bills, and that there is no reason why those in areas where there should be an income–expenditure deficit, on its reckoning, should pay towards areas where local incomes should more than be enough to deal with these liabilities. What might the result be?

In some areas, local administrators would take the view that in order to avoid raising local income taxes to the point where the administration would be unelectable, it would be necessary to pay the social security bill by cutting back on road improvement programmes and renovating school buildings, to ensure sufficient resources to finance the element of the social security bill unmet by central government. In these areas, roads would soon fall apart, and the schools would deteriorate rapidly. Many local people would react by refusing to pay their local taxes, resulting in a cash crisis for the local administration, or an explosion in the local jail population. Others would move away to areas still receiving central subsidy. In time, these areas would also suffer central cuts, as their notional and actual local incomes rose.

Alternatively, a 'blame culture' would develop in areas suffering central resource clawback. Cuts in local services would be blamed on the 'workshy and feckless'. One can imagine the atmosphere of societal tension and distrust which might result. Hopefully, it will never happen; the consequences of such a policy would be unacceptable in terms of societal cohesion and deliverability.

So why is it acceptable in the field of housing benefit, which is a part of the national social security system? It is less general, but does that justify it?

Back to the reforms. The government is intent on introducing a form of standard local housing allowance to the council sector. This essentially will entail setting rent rebates at average rent levels in the council area by property type and size, rather than on actual rent levels, as at present, and allowing tenants to keep any difference between the allowance and actual rent. If the allowance is lower than the actual rent, the claimant would have to make up the difference from his or her own pocket. This would theoretically encourage tenants to consider carefully where they should be living to get value-for-money – OK, perhaps, where there is a genuine market in social housing and real scope for mobility. Not OK, perhaps, where this would lead to severe overcrowding in an already overcrowded sector, as tenants scramble to stay within or below allowance levels by down-marketing to smaller or less desirable properties, leaving the larger ones for those able to afford the full rent. In the case of rent rebates, if still deducted at source, 'keeping' a rent rebate would probably mean a credit to the tenants' personal account.

A collateral reform proposed is to abolish rent rebating as such, and paying housing benefit to tenants as in other sectors. This is part of the 'Choice and Responsibility' agenda of the Blair Labour administration. The argument runs that housing benefit is just as much a part of personal income as Job Seekers' Allowance or tax credits, and it is up to the recipient to decide how to use that benefit. Rent rebating, which is a deduction from the rent rather than an amount given, vetoes this choice, and therefore offers no budgetary choice to the claimant whatever. Giving council tenants the equivalent to the deduction would be to treat them just as any other benefit recipient, and level the playing field with those getting private sector housing benefit including other social housing tenants with RSL landlords.

Why, though, shouldn't council tenants be treated the same as other social housing tenants? To retain direct deductions in the form of rent rebates would surely be to treat them as if they could not be trusted to pay over their benefit to meet rent obligations. Many council tenants would surely welcome the freedom to receive cash instead of a reduction, and make their own decision as to how to apply it.

It happened in 1981. It resulted in local authorities asking the then DHSS to reverse the decision, as arrears mounted, along with evictions for non-payment; and even the tenants' lobby supported the return to direct deduction, partly on the grounds that service levels would inevitably suffer through forced economies as a direct result of lost income. Perhaps things will be different now. The risk of such a policy is that arrears will mount: those in receipt of rent rebates often have very little money to spend on necessities, and have to make hard decisions on what to do with their limited resources. Removing the difficult decision as to whether to apply housing benefit to paying rent or buying food or children's shoes is in fact a relief for many households. Admittedly, the reform will not apply to 'vulnerable' tenants or those in arrears, but there is the serious collateral risk of reduced HRA income, and personal financial difficulty, even culminating in the loss of the home and sentencing to temporary housing as a result of a decision of intentional homelessness. It's a hard one to call. Perhaps the best thing to do would be to 'pilot' the reform in a limited number of councils, to see what the effect on HRA income and arrears levels is, along with the number of Notices of Seeking Possession issued in respect of non-payment.

There is in fact every justification for keeping the rent rebate system as it was in 2004, on the grounds of affordability. Rents are set according to rent restructuring principles. They are supposed to be set on the basis of average manual income levels (70% of the formulation), and the value of the properties let (30%). If someone is earning below manual earnings levels, and mathematically there must be such cases, then by definition such a person could not be expected to afford the rent. It is therefore consistent with the notion of affordability implied that the rent they should be expected to pay ought to be adjusted accordingly. It is therefore reasonable to reduce the rent payable to take account of the less than average income. This is what rent rebating does, and it was thought up many years before rent restructuring – prescience indeed! So why spoil a good thing, which is consistent with rent restructuring policy, by moving across to a system where tenants are liable for the full rent, albeit being given some financial assistance to pay it?

Rent allowances are housing benefit (private sector) and are cash, cheque or direct account payments to households in private rented or housing association accommodation whose incomes and capital resources fall within the relevant criteria for the housing benefit system. Tenants who are 8 weeks or more in arrears forfeit the right to this personal payment, which will go instead to the landlord as housing benefit direct. In 2004, housing benefit was assessed on the level of rent assessed by the rent officer service to be eligible on a case-by-case determination basis. The portion of rent above the rent officer threshold would not be eligible for support, and would have to be paid by the claimant from other resources. This limitation was thought to provide an incentive to tenants to negotiate the rent level at the rent officer-assessed level.

Prior to 1989, private 'regulated' *1977 Rent Act* tenants could in any case apply to the rent officer for a 'fair rent' which would be the limit of rent chargeable on a two-year review basis. This was swept away in the wake of the assured tenancy regime for new housing associations and private tenants with on-resident landlords as a result of the *1988 Housing Act.* The restriction of housing benefit at the level of the rent officer determination, which took into account comparable market rents for that form of property in the determination area, might have been seen by some idealists as a lever for tenants to get rents down to a reasonable level; but in which case, why should landlords let to housing benefit cases at all? In practice private tenants paid the difference, or sought something cheaper.

The 'choice and responsibility reforms' following the DWP's '*Building Choice and Responsibility*' proposals published in October 2002 aimed at sweeping away property-specific housing benefit, and paying at average rent levels for that sort of property in a defined geographical area, subject to a means test. Direct payment of housing benefit to landlords, which was something that could be done with the agreement of the tenant, would be swept away, except in cases of vulnerability and serious rent arrears. It was thought that this could be introduced more rapidly for the private and housing association sectors, due to there being a larger market for these properties – more choice available to tenants, partly as a result of the introduction of choice based lettings systems in the housing association sector from 2002 onwards. In standard cases, local housing allowances would be paid at the average rental rate. Tenants paying less than this level would be able to keep the difference, incentivising choice of cheaper properties, and tenants paying more would have to find the difference from their own pockets, again incentivising choice. Either way, the state would be able to more nearly predict the need to spend on this very volatile element of AME.

To test the proposition, several 'pathfinder' authorities were chosen to implement the reform in the private sector initially, to see if it would work. At the time of writing, the jury is out on the effectiveness of such measures, and full-scale implementation is not in place. Some housing associations have tested out the abolition of housing benefit direct to landlords: one, London and Quadrant Housing Trust, a housing association with properties mainly in South London, saw arrears rise from 3 to 8% of the annual rent roll in six short months after introduction, which may be indicative of the wider impact (2003/04 experiment).

It will be interesting to see how enthusiastic lenders to housing associations will be about these reforms. Housing associations rely upon stable and predictable rent roll income to service development loan liabilities, as well as to run housing management and maintenance services. Anything which increases the likelihood of arrears increase and thus income reduction has to be of concern to housing associations and their lenders, who expect prompt repayments. If housing associations cease to be good bets for loans, then institutional lenders will look elsewhere: the market for loans is very large. If the HB reforms cause lenders to pull out of the RSL market, the government's *Sustainable Communities Plan*, which involves the production of a large number of for-rent social housing as part of the package to increase the quantity of affordable housing in high-demand areas, will be threatened. Private developers building for rent will face a similar problem. Is this joined-up thinking?

Time will tell as to whether these reforms are carried through in the social rented sector, but caution should be exercised on the basis of the early experimentation in the social rented sector.

Other options for reform

If rents are set above the level which households can afford, and there is a social commitment to ensuring that people who are on lower incomes have a home, and have a choice as to where to obtain it, then there is a case for the availability of assistance in the form of housing support. As long as governments refuse to return to the regulation of the private and voluntary rented sectors, then we seem to be stuck with it, and the only question is how that support should be calculated and paid.

Alternatively, why not return to rent regulation? One of the arguments advanced by William Waldegrave, the Conservative Housing Minister who presided over the abolition of *1977 Rent Act* regulation by bringing in the *1988 Housing Act* assured tenancy deregulated regime, was that rent regulation was demonstrably reducing the size of the private rented sector, and creating homelessness, as landlords de-invested in the sector and put their money elsewhere. He thought that deregulation would revive the fortunes of the private rented sector, which in 1988 stood at around 10% compared to 7% in 2004. Landlords would offer properties at an asking price determined by their perception of the market comparables, and rents would be set on the basis of a bargain struck with willing effective consumers. Rents would naturally rise and fall with supply–demand conditions.

Waldegrave thus paved the way for vastly increased volatility in the government's current spending account (now AME). Housing benefit requirement would now in principle be largely unpredictable, as it would be determined largely by bargains struck in the open market, unless it was capped, which would be tantamount to central intervention in a free market, an anathema to the Tories of the day. This sort of intervention did in fact follow, under David Curry in the early 1990s, where there were early attempts to try to cap private sector housing benefits at levels to encourage bargaining by tenants designed to drive rents down. They forgot that there are plenty of would-be tenants who can afford the going rate without the odd negative incentive of not having enough housing benefit money to bid to the full asking price. There is no law which outlaws discrimination against housing benefit claimants when deciding which tenant to take on.

* * *

An understanding of the various interlinked elements of housing finance, and its constraints, is essential if the twists and turns of general housing policy are to be adequately grasped.

References

Beswick, K. and Bloodworth, J. (2003). *Risk Management Topic no. 4: Risk Mapping-Dilemmas and Solutions*. Housing Corporation.

Department of Trade, Environment and the Regions (DETR) (2000). *A New Financial Framework for Local Authority Housing: Resource Accounting in the Housing Revenue Account.* HMSO.

Department of Trade, Environment and the Regions (DETR) (2001). *Guide to Social Reform.* HMSO.

Housing Corporation (2000). *Effective Risk and Business Management.*

Housing Corporation (2003). *Mergers, due diligence and housing associations – a good practice guide.*

ODPM (2003a). *The Local Authorities (Capital Finance and Accounting) Regulations (England) Regulations 2003.* HMSO.

ODPM (2003b). *Housing Revenue Account Manual.* HMSO. See also: www.housing.odpm.gov.uk/local/hsg/hram

ODPM/Building Research Establishment Division (2003). *Estimation of the Need to Spend on maintenance and management in the Local Authority housing stock.* HMSO.

ODPM (2004). *Local Authority Projects Endorsed by Interdepartmental Project Review Group (PRG).* HMSO (latest list can be found via the ODPM website).

Zebedee, J. and Ward, M. (2004). *Guide to Housing Benefit and Council Tax Benefit.* Shelter/Chartered Institute of Housing.

Bibliography

CIH (2003). *Supporting People; Good Practice Briefing no. 25.*

Garnett, D. and Perry, J. (2004). *Housing Finance*, third edition. CIH.

Gibb, K., Munro, M., and Satsangi, M. (1999). *Housing Finance in the UK. An Introduction*, Second edn. Macmillan.

Kemp, P.A. (2000). *'Shopping Incentives' and Housing Benefit reform.* CIH.

King, P. (2001). *Understanding Housing Finance.* Routledge.

Lindelow, M. (2002). *Holding Governments to Account: Public expenditure analysis for advocacy.* Save the Children.

ONS/ODPM (2003). *Local Authority Revenue Expenditure and Financing: 2001/02 Out-turn* (or latest relevant year). HMSO.

ONS/ODPM (2004). *Local Government Finance Key Facts: England*. HMSO. Latest downloads can be obtained from www.homeoffice.gov.uk

Smith, C. (latest edition). *Housing Benefit for Housing Managers in the Social Sector.* Russell Press.

Websites

www.odpm.gov.uk
www.hm-treasury.co.uk
www.cipfa.org.uk
www.cih.org

4 Housing development

Introduction

Social housing development is now the preserve of housing associations and some private developers, but it wasn't always that way. The majority of social homes in the UK are still in the ownership of the councils which built them – some 3.54 million homes in 2002, or 13.8% of all homes in the UK (ODPM, 2003). As indicated in Chapter 1, the social housing development programme (e.g. of around 20,000 homes a year for rent in England (at 2004/05)) comes nowhere near meeting housing need, and the consequences are housing stress and homelessness. There is a good case for boosting housing production for rent, and perhaps for low-cost home ownership, and in recent years the government has increased its commitment to moving nearer to enabling the production of social housing on a more realistic scale, but there is a long way to go.

This chapter will examine a number of interlinked themes. It will seek to explain why RSLs are the main producers of social housing, and how this situation has developed. It will examine the role of councils as strategic planners in estimating housing needs requirements, enablers of development through the planning system, and their emphasis on regeneration of areas, and major works and improvements to stock, a complementary role to that of new housing production. It will explore the pivotal role of central government and non-governmental organisations, principally the Housing Corporation, as major sources of development resources, working increasingly with private financiers to provide and enable funds for the development of new housing, and the regeneration and conversion of older stock. This will be set firmly in the context of the *2003 Sustainable Communities Plan*, as amended, which stresses the importance of adequate affordable housing provision in the growth areas, focusing on London and the South East, and the development challenges this will entail for all concerned (see Chapter 1).

In the light of the under-supply of social housing, and the consequent reduction in turnover in the housing stock available, social homes will have to accommodate the same household for far longer than previously, and therefore the imperative of developing homes to meet the changing liveability requirements implied by life-cycle changes is ever more important.

The enabling and production of new build development for rent and for sale

Social housing development is a process which requires the estimation of the amount and type of housing needed at various geographical levels, ranging from districts through counties, regions, countries and even continents. Some of these aspects have been discussed in Chapter 1. Having estimated quantitative and qualitative requirements in the short, medium and long term, and estimated the extent to which the private sector can provide homes in the open market to meet effective demand, it is necessary to consider the cost of providing for the remaining households, actual and projected, who may have to look to the non-market sector for housing, against more general macroeconomic considerations.

It is necessary, at this stage, to identify how much land is required in the right areas to meet the implied targets, in relation to development land availability, as informed partly but not wholly through the planning system, and bring forward additional land through town and country planning policy if there is insufficient designated for residential development.

It is then essential to determine the most cost-effective way of supplying good quality housing, from the viewpoint of design, to ensure that properties are durable and designed flexibly to ensure long-term occupancy through life-cycle changes and are sensitive to user requirements, taking into account the possibility of converting non-residential buildings, and density considerations.

The next step involves ensuring that there are sufficient agencies around to develop the properties required, and to manage them when completed. Only then can development programmes be commenced in a rational manner.

This is the theory, and just an outline of the process. The reality is far less orderly, and fraught with political issues, as well as capacity and technical challenges. This is what makes housing development such an interesting and sometimes risky career choice, but there is a certain joy at seeing the results of all this planning in new housing development which meets need and meets with resident satisfaction.

The strategic planning process

How many homes are needed?

Every local authority has a duty, under the *1985 Housing Act*, amended by the *1996 Housing Act*, to estimate the need for affordable housing in its area by conducting a housing needs survey. This should take account of the formation of new households as well as in- and out-migration, and the degree to which the private sector – rented and owner-occupied – can meet predicted housing requirements. The 'housing requirement' of an area is the total amount of housing needed by households in an area at a given point in time, projected into the future on the basis of household formation and migration trends.

A basic formula for determining how many new homes are needed, affordable or otherwise, may be outlined as follows:

> **INFLOW** (new households forming **plus** in-migrating households) **minus OUTFLOW** (out-migrating households **plus** households no longer requiring housing)

This would give a crude measure of the number of additional homes needed on the basis of natural growth, natural 'wastage' through death; inflows and outflows.

The formula can be used in respect of different sizes and compositions of household in order to estimate the need for different sizes and types of housing. Factors which need to be included, to estimate property-specific requirements include:

1. *Demographic profile*. If a population is ageing, then there will be a requirement for more smaller properties, perhaps with one and two bedrooms, which are built to mobility and wheelchair standards, and possibly more sheltered housing schemes. If it is expected that there will be a growth in the number of younger households, through migration or natural growth, then it will be necessary to prioritise the enabling of more family accommodation to meet the requirement.

   Increasing birth rates, combined with increasing household formation and in-migration, implies a need for more family-sized accommodation.

   Average life expectancy is still increasing, which means that, all other things being equal, death rates are decreasing. This means that there are fewer vacancies arising in existing housing through death, which in itself reduces 'second hand' housing supply, and increases the need for housing suitable for the old and very old.

   It is essential to factor the demographics of migration to housing requirement estimates. An area might be particularly attractive to older people as a retirement destination, due to climatic, scenic or amenity factors.

2. *Economic profile and turnover rate*. If the area is subject to economic decline, this might imply a reduction in demand for housing due to out-migration, and an emphasis on the

enabling of social housing through housing associations, depending on the turnover (vacancy rate) in existing social housing. High vacancy rates will reduce the requirement for new housing, although this will be property-specific, as younger households are more mobile, and can be expected to migrate out of the area if there is somewhere else within moving distance where economic conditions are better.

3. *Social factors*. This is allied to demographic factors, but focuses more on change in household size and numbers through variations over space and time in the age of new household formation, of the creation of households which wish to cease to be 'concealed' (i.e. living with another household, as in the case of a young couple living with one or other sets of parents), and of the increase in household numbers through increasing divorce and separation trends.

This may give an approximation to the total housing requirement in an area over a given period of time. Not all the properties will be affordable to those on lower incomes, and it is necessary to try to estimate the number of affordable homes required by taking account of the economic characteristics of households likely to require new, second-hand, or converted property.

Affordable housing requirement

There are a number of factors which need to be considered in order to estimate the affordable housing requirement. The first are **demand** factors, which generate a gross requirement figure.

1. *Exclusion from owner-occupation (income:property price ratio)*. If incomes are low in relation to specific property types, for example, family-sized houses, then this implies that a number of new or existing incoming households will require housing at lower than market rates.

   The number of new and existing households whose requirements cannot be met through the owner-occupier market can be assessed by comparing average incomes, multiplied by standard lender multiples, plus an assumption about average savings levels, to the average cost of property plus other acquisition and moving costs, broken down into bed sizes. Trends deduced from the previous stages can be used to deduce the size of the total requirement, and house-price trends can be used to estimate the present and future average cost of owner-occupied housing for the period desired. An estimate of number of households requiring affordable or 'sub-market' accommodation from a consideration of *lack of ability to buy* could then be worked out by deducting the number of households who meet the income multiple and savings criteria in relation to property-specific house prices from the total 'inflow' comprising newly forming households and those arriving from elsewhere. The key word here is 'estimate' – such a predictive model is only respectable if there is a statistical factor – a margin of error built in, which produces a band containing probable outcomes rather than a single figure.

To make the model dynamic, assumptions need to be imported about real income growth over the period, and real rises in property prices in respect of different house sizes. Trends on the former can be obtained from *Social Statistics* reports issued by the Office of National Statistics and through local survey information, and on the latter by examining property price trends by quarter issued by *HM Land Registry*, or again, through local survey.

2. *Exclusion from private renting.* It may be assumed that those excluded from owner-occupation may seek private rented accommodation, which may be cheaper when housing benefit availability is taken into consideration. Armed with a range of income and savings figures taken from households in the 'excluded from owner occupation' category, it should be possible to estimate how many of these households could potentially afford to rent at prevailing market rentals for suitable types of rented property by type, size and location. The remainder is the number of households who cannot access owner-occupied property at prevailing market rates, and

who potentially have to look to sub-market or 'affordable' housing to meet their needs.

The first part of the estimation involves guessing demand without reference to supply. It cannot be assumed that the market will satisfy the demand from all of those who can afford to buy or rent. If it can, then the number of households requiring sub-market housing is simply the total number of households estimated to require housing minus the number who cannot afford to do so, either by buying or renting, projected over the number of years required for planning. If there were one hundred additional households expected to require housing in an area, and fifty of them could potentially afford to buy or rent at market levels, but there were no additional houses or flats to buy or rent, either through new production or turnover, then none of them would be able to satisfy their housing requirement in the area.

It is therefore necessary to estimate housing supply over the plan period, in order to find out what the net affordable housing requirement is likely to be, to inform the size of affordable development programmes. This can be done in several stages.

1. *Numbers which will be produced.* In the short term, how many new properties of various types and sizes have been given planning approval, and how many conversions, which will become available over the plan period? This is relatively easy to state for the first year, and more speculative beyond that. A rough estimate can be given by considering the amount of land available classed for residential land use, and working out how many dwellings could be built on it at permitted densities over the period. From this potential maximum figure would be deducted a figure based on actual annual production levels over the past few years, assuming that there are no sudden booms or downturns in the construction market, unless the trends in these areas are clear, in which case this non-linearity can be factored in. This would give a reasonable estimate of new and converted homes producable in the area.

2. *Price of homes produced.* To each of these homes must be attached an average price, based on current Land Registry figures, averaged by bed size, and relativised to location. It is in practice hard to predict house-price trends into the future with any certainty, and a straight-line projection may not capture the complexity of trend. Holding the increase factor constant would not be particularly helpful in light of past non-linearities. If this can be done, supply can be assessed as the number of homes produced at given prices. Most will be for sale, some for rent, and a judgement needs to be made as to the division, perhaps by looking at the actual division over the past few years in respect of new developments in the area.

The affordable housing requirement quantum

It should now be possible to estimate the affordable housing requirement by taking estimated demand from supply. A negative figure indicates the estimated level of shortage of affordable accommodation in the area, which may increase or decrease over the plan period depending on estimated trends fed into the model. That shortage can also be specified in terms of property mix, and location within the area, depending on the complexity of the model. A positive figure indicates that there is no requirement to produce or enable affordable housing in the area.

Planning policy implications

If it turns out that there is currently, or is likely to be, a housing shortage in the area – that is, that estimated supply falls short of the estimated requirement – then it may be necessary to bring forward more land by change of use-classes, for residential development, or to modify policy on density within the limits of legislation and good practice in town and country planning.

If it turns out that there is likely to be a deficit of affordable housing, planning policy may come into its own by examining carefully the scope to attach planning conditions to residential applications by requiring a given percentage of those properties to be affordable, or though greater and more effective use of *Section 106* of the *Town and Country Planning Act 1990* (which modified *Section 52* of the *Town and Country Planning Act 1971*) to induce developers to give land or housing to a third party (e.g. a housing association) to ensure that the community shares in the profits made by the developer. This is especially feasible where a change of land use is requested by a residential developer, or where it is envisaged that the development might not quite meet the density standards required in the local plan.

At the end of the estimation process, and taking local planning policy changes or flexibility within the existing framework into account, it should be possible to come to a view as to how many affordable homes, of what sort, and within which areas, are needed to meet estimated need. It is then a matter of trying to decide the form of the affordable housing in terms of tenure: and this goes beyond town and country planning and into the realm of mobilising relevant agencies to produce the goods, and it is to this that we now turn.

Who can produce affordable housing? Why not councils?

Local authorities can build houses. They can borrow from the Public Works Loan Board, a division of the Treasury, over thirty years for improvement or conversion work to existing property, and over 60 years for new build. Many can also borrow from the private sector, if judged efficient and effective enough by central government under the Best Value inspection process. Loans can be advanced to the council's general fund. They can pay the loan instalments through rental income, and receive loan debt subsidy as part of their HRA subsidy settlement. They can also use 25% of housing and 50% of land capital receipts to supplement their borrowing. The only constraints on them are the prudential borrowing limits, which are subject to regional cash limit, as previously discussed. It might be thought that councils would readily take up this opportunity, especially

Fig. 4.1 Council houses, Surrey.

Fig. 4.2 Council flats, Surrey.

Fig. 4.3 General needs 1930s council flats, Surrey.

Fig. 4.4 1960s council sheltered housing scheme, Surrey.

since the demise of Local Authority Social Housing Grant, which would effectively have allowed them to get access to two social homes for rent rather than one, due to the grant: development cost ratio being around 1:2. They are the planning authority within which they operate, and surely the local authority would give itself preference for planning permission over all-comers if it came to a fight. And in any case, the local authority housing department, being part of the same council, could be expected to have a very detailed knowledge of the local planning regulations. Add to this the fact that many councils have land they can develop on, and therefore do not need to spend scarce financial resources on acquiring an asset which often accounts for one third or more of the total cost of developing a property, and it would seem that council house development should be very cost-effective. Consider also that rents do not have to be set to cover the development loan element not covered by subsidy, since a significant proportion of council housing was built over 60 years ago, and therefore those rents are no longer paying off debt charges, and can perhaps cross-subsidise the loan-related costs attached to newly built homes.

All of this is true, but a combination of parsimonious HIP settlements up to the 2004 Spending Review, the restriction of capital receipts expenditure apart from a few years of 'holidays' on constraints, the advantageous nature of LASHG, the pressing need to deal with the major repairs backlog, and latterly, the injunction to ensure that homes are decent by 2010, have caused most councils to abandon their historic role as developers, and leave it up to housing associations, who have greater financial freedoms, and after all serve more or less the same client base.

This position is in many senses regrettable. Local authorities are democratic bodies, with councillors directly accountable to all enfranchised local people through the ballot box, which provides their only legitimacy. The provision of housing for local people to meet neighbourhood need is surely one of the most poignant expressions of the local state responding to need within its boundaries. Councillors have an intimate knowledge of not only their electorate through surgeries and walkabouts, but also care about their local areas in terms of environment. Social infrastructure and economic wellbeing, or at least have to make a pretty good show of these things when standing for election to represent wards or districts. If they get decisions wrong, they know that they have to face the local electorate in their communities. For all of these reasons, councils are certainly appropriate bodies to develop housing, as well as to manage and maintain them, and it is only dogma which stands in the way of their building again.

Look at the competition. Housing associations often span entire regions, are governed by boards, drawn from selected members with absolutely no link to any democratic election process. Their paid officers advise the board, and carry the day in policy terms due in many cases to superior knowledge of the sector, and the fact that chief executives are board members as well, and can be expected to exert their authority. They are overseen by an unelected non-governmental organisation, the Housing Corporation, which is responsible to parliament and not any government department, run by a selected board of political appointees.

Admittedly, many housing associations started out as locally based bodies, often charitable and well-motivated, set up to try to meet aspects of local housing need, and some still are – witness the number of associations with 'church' or 'churches' tagged onto their names, which started as organisations to do something positive about their congregations' concerns about the state of the local housing situation – but many have grown far beyond these roots. They argue over nominations rights, often remaining fiercely independent of the local councils in which areas they develop, as independently constituted organisations.

Regional housing associations must balance the needs of one area over another, and cannot be expected to operate in areas contiguous to councils, or to prefer one to the other. Even LSVT associations, which started life as recipient of council stock, and came with a virtually complete transfer of staff deeply wedded to council ways, and managed by boards including a goodly sprinkling of local councillors, have now spread their wings and often operate far out of their original geographical areas of operation. Many have shed their original names which identified them as transfer organisations – witness the transformation from Swale Housing Group to the commercialistic AMICUS. What council did Ridge Hill HA take its stock from? Where does the stock of Hereward Housing come from? It is true that many of these bodies are tied contractually to their parent authorities, in discharging statutory functions and the like, but they are fast becoming not-for-profit organisations with an independent ethos.

It is political ideology which has separated the housing development function from local authorities, and the result has been an unnecessary and regrettable complication of the whole process, which creaks with inefficiency, is needlessly expensive due to the duplication of functions and burgeoning of organisations with fingers in a pie of insufficient size to do more than provide a starter.

Many arguments have been produced over the years to weaken the case for council building in preference to other developers, and they are all contingent or bogus. Here are some:

1. *Councils are inefficient spenders of public resources, lacking the profit motive and discipline of competition of developers to produce homes at a keen price. Leave development to business or business-like organisations.*

This was the argument advanced by the 1979 Conservative Thatcher administration and the Major administration after it, and not contradicted by any New Labour administration from 1997 onwards. It is inherently flawed. A commercial organisation will seek to maximise its profit, in order to pay its shareholders, by selling for as much as it can or by producing things or providing services as cheaply as possible, or both. Naturally, the product has to be marketable, otherwise nobody would buy it at all, or they would run to another supplier, which is where the smoke and mirrors of marketing and advertising come in as well as adherence to minimum standards. When the product is social housing, for the poorest who have little choice but to rent at a relatively low price, rather than executive four or five bedroom detached houses in Dorking, South Woodham Ferrers, Altrincham, Henley-on-Thames or even on the shores of

Loch Lomond, this is a marvellous opportunity to knock out relatively cheap housing. After all, developers know that they will not be able to unload their wares for more than a very average price to a local authority or housing association, due to their cash constraints, and they certainly would not build for social rent, because they would never be able to meet their development loan obligations on the basis of the rent they could get away with. They would, however, expect to make a similar profit on the deal as they would with any other. Economy in the quality of building materials used, rate of labour pay, and standard building types without reference to client requirement sensitivities, should not be confused with market-facing sensitivity.

Another fundamental issue is that developers are for-profit organisations. A profit is not the same thing as a surplus, to be ploughed back into production. Profits are what are taken out of the cashflow of a business as an investment reward, and are precisely what are not ploughed back into the business. The need for private enterprise to make a profit is both a constraint on quality and a source of needless appropriation of public money. Why should the public line the pockets of a developer or company shareholders, when it could apply the same money to the creation of social homes, road repairs or virtually any other social product which benefits most or everyone? There is nothing wrong with private enterprise in its place – it is, after all, the source of tremendous motivation and wealth, a portion of which is channelled into the taxation pot which helps pay for much-needed public services. It is questionable as to whether state resources should be used to reward private investors on the back of providing for the poorest, when the surplus could have been use to fund more public provision, when there is an efficient alternative; and there is.

Instead of turning development over to the private sector, it would have been sufficient to tighten up on council financial probity, procurement and operational efficiency, but political directionality precluded this outcome. Now we have councils which are efficient and effective operators by any standards, being awarded three stars by the Audit Commission, having HRA Business Plans which are accredited as Fit for Purpose, councils with borrowing

Fig. 4.5 1960s tower block, London.

Fig. 4.6 1960s tower block, London.

Fig. 4.7 Deck access 1950s medium rise flats, London.

Fig. 4.8 Deck access 1960s medium rise flats, London.

freedoms awarded after a through scrutiny of their financial management prudence. Add to this the fact that they are not-for-profit operators. Are these not extremely strong candidates for being social housing developers if any organisations are: so what's stopping a return to council housing development?

2. *Councils built the most appalling homes in the 1950s through to the 1970s. Do we really want to turn over development to organisations which produced the sort of tower blocks which have been blown away in Hackney, and the concrete jungles ruining the edges of too many cities with their anonymity and sheer ugliness? They have demonstrated that they can only produce mediocre rubbish: they are not fit to develop.*

Granted, there are examples of appalling housing built by councils across the country, and in municipal enclaves across the world. There are new towns with no heart or soul where gangland territory boundaries are marked by spay-painted symbols, tower blocks where the lifts work sporadically, and whose common parts are smeared with graffiti, where inadequate CCTV monitoring or caretaking increases the risk of mugging or worse. There are estates where only crack-heads venture out at night, and where the experimental building methods – the 'modern methods of construction' of the 1960s and 1970s – have created whole blocks of deserted, boarded-up slums waiting for the bulldozer. Granted, much of the council development domain does not look good.

It has to be asked, why are these places so appalling? Who designed and built them in the first place, and who compromised on quality standards which led to such a dreadful pass? Take tower blocks, the symbol of all that is worst about council building. Originally, they were conceived as villages in the sky, after the experimental work of *Le Corbusier* in the 1920s and 1930s. They were to have residential zones, floors dedicated to shopping, work and recreation, and any number of other functions to be found in villages and towns – except that they would be provided vertically rather than horizontally. That was the idea anyway. What we got was home after home piled one on top of the other,

Any developer is capable of building tat, some of which is redeemable, so perhaps all development should be banned absolutely and for all time, just in case they all produce rubbish!

Faceless developments also abound, courtesy of the speculative builders of the 1930s, on the edge of almost every English town of any size. Bungaloid growths attach themselves to the arterial routes out of major cities like a cancer: rows upon rows of identikit semis, you could be anywhere!

Enough has been said to establish the point that anywhere and nowhere estates are not the preserve of council building, although local authorities have not helped themselves in this respect. At the end of the day, they were clients of architects and developers, and only so much can be laid at the door of forced economy.

The estates and buildings of the past represent mindsets of the past, and to a certain extent we have to live with that, unless we can convert or improve the dwellings to modern standards. For example, take the 'traditional' estate with its low-rise terraces and semis, greens and neighbourhood shopping areas: most towns and cities have one or two. Typically they were built between the 1920s and 1960s. If you look at them carefully, they have a number of features in common which reflect the life and times of the day.

The houses often have two receptions and small kitchens, reflecting a time when there were fewer kitchen appliances in use, perhaps a stove and maybe a refrigerator, when the family would congregate in the dining room for meals and the lo[u]nge for social activities. There may be a box room upstai[rs] far too small to take the high tech equipment and paraph[er]nalia which every child expects nowadays. The bedroo[ms] may be smaller than ideal; again, due to the fact that ther[e w]ere fewer possessions to cram into them. Going outside, i[t is] unlikely that there is adequate space for parking one o[r tw]o cars in a drive, and the garden may be put to this pur[po]se, as well as the road outside. Garages may be located [in] blocks round the corner, and may well be under-used, [du]e to the fear of vandalism, as they are more than likely t[o be] out of sight of dwellings, and surveys such as that supp[orti]ng the *Institute of Housing* publication

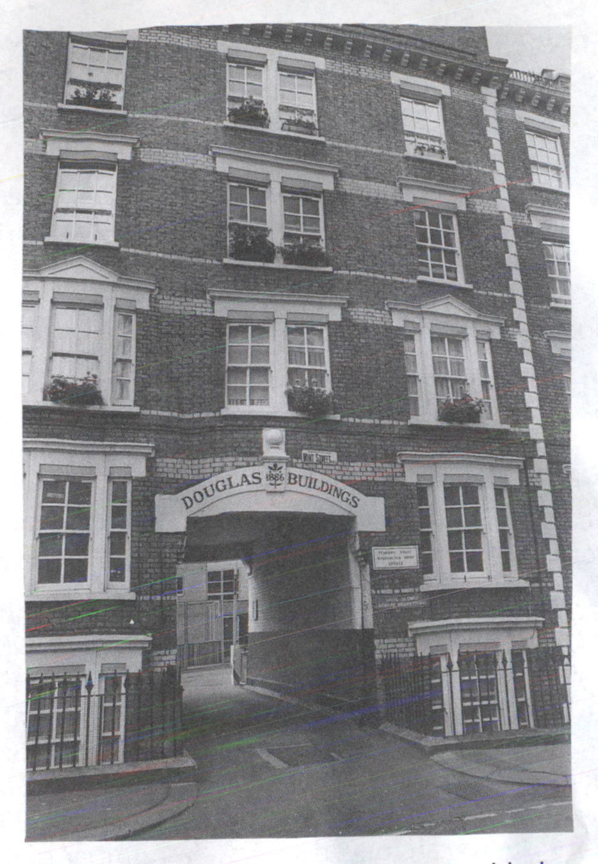

Fig. 4.9 Refurbished Victorian council mansion block.

Fig. 4.10 Refurbished Victorian council mar[illegible] block.

Fig. 4.11 Council low-rise housing – regeneration area.

Fig. 4.12 Council infill, regeneration area.

Homes for The Future (1991) indicate that car owners prefer to be able to see their vehicles from their homes.

Moving outside the home, there may be a green area, originally either for ornamental purposes or for walking on, or for local kids to play ball games – although this use is often prohibited. They were designed into estates in the days when television ownership was low, and when it was considered good social engineering to encourage healthy outdoor pursuits by making greens available. It may also have been placed there to encourage or facilitate collective activities, such as neighbourhood events and for people to commune and pass away the time. Society has changed, and these features are too often neglected relics of past mores and aspirations.

Further on, there is a parade of neighbourhood shops – government design bulletins encouraged councils to include commercial and social infrastructure, such as shops, health centres, churches, and even pubs, into estates, to serve relatively immobile local communities without significant access to mass transport. Today the shops are still there, but one is boarded up, another is a 24/7 convenience store selling anything and everything, and yet another is a charity shop. The butcher and baker moved out in 1987 when a major retail chain developed their edge-of-town supermarket a mile down the road. The pub still enjoys reasonable custom, but it has changed hands several times, and is less popular than it used to be when fewer residents owned cars. The health centre now trebles as a district housing office and social services surgery.

When it opened in 1935, the parade was well used on a daily basis: no-one had a car, the bus service to town was only marginally better than it is today, and supermarkets with relatively cheap prices due to economies of bulk purchase were unknown. The health centre was welcome, in days when proprietary medicines were far more restricted, when diets were poorer, and health education was far more limited.

Driving around the estate, one is struck by the narrowness of the roads and the smallness of the turning circles and hammerheads, reflecting days when cars were smaller. And

soaring above the estate in a hot-air balloon, we cannot fail to notice that the designers have clustered the entire estate around its community facilities, trying to create a community in bricks and mortar.

The balloon drifts across to the inner city. Down there are ranks of Victorian terraces, now gentrified; but one hundred years ago, it was a different story. The balloon drifts through a time-cloud, and lands in the Victorian street, with its inadequate sewerage, and houses in very poor repair. Mass housing at the beginning of the twentieth century grew out of a concern to provide healthy housing to replace these slums, often because the middle and upper classes in power feared the spread of disease from these places to their own residences.

The central point is that that the developments of the past grew out of past concerns, and that we cannot meaningfully judge those who enabled them by present standards.

Councils would not build in the manner they used to if they built again, because the challenges are different. Today, the concerns are more to do with sustainable communities, to try to ensure that there is a tenure mix and that there is a reasonable mix of ages and household types. Homes would be energy-efficient, conforming with contemporary standards. They would probably be built to lifetime standards, so that households could continue to get full use of the dwellings throughout their life-cycle, and that no household would be disabled due to the design of the house or flat. There would be sufficient space to accommodate the parking of two or three cars per dwelling, or far fewer, depending on the degree to which the council wishes to wear its green credentials on its sleeve through urban design and planning. (For a short discussion of the lifetime homes concept, see: Habinteg Housing Association (2004)).

In short, it would be wrong to judge today's councils by the products of past generations, because the social and economic scene has shifted. Today's councils could make at least a good a job of building decent homes for those who cannot afford to compete in the marketplace than others, and without the requirement to make a profit, would not be tempted to cut corners in the interests of making a quick buck.

Fig. 4.13 Housing Association 1990s sheltered scheme.

Fig. 4.14 Housing Association 2003 – general needs flats.

Fig. 4.15 Housing Association 2003 – sheltered housing complex.

Fig. 4.16 Housing Association leasehold scheme for the elderly under construction, 2004.

3. *Housing associations have, since the disciplines of the fixed-grant private finance regime following the 1988 Housing Act, become extremely efficient builders. They are able to match private finance with public subsidy pound-for-pound in producing homes for rent and low-cost sale, and produce excellent homes within the relevant Total Cost Indicators, without draining the exchequer in the process. There is therefore no need to engage councils in the business of building, rather than enabling through planning and land provision. Additionally, with the increasing spread of common housing registers, common allocations policies and regional choice-based lettings policies, combined with regional and sub-regional nominations arrangements, it would be a needless duplication of effort for there to be more than one type of social housing developer, since all the customers are in common. Leave social housing development to the housing association sector.*

This argument is fundamentally flawed, for a number of reasons, and contradicts the principles of best value, which have run through the social housing sector since the late 1990s. If local authorities were able to develop housing, there would be a direct comparison between the council and housing association process and product, and a judgement could then be made on a case-by-case basis on the relevant efficiency and effectiveness of each organisation as a social housing developer. In some areas, housing associations may be superior in many respects, using better procurement processes, with tighter contract-management, and producing homes which more nearly meet the aspirations of ultimate customers than the council, in which case best value considerations would lead to the preferment of housing associations as developers there: but the reverse is equally possible. The fact that councils are no longer developing pre-judges the issue: all that can be done is to compare past council housing performance with present housing association activity (councils would win hands-down on volume), which is no basis for a rational comparison, due to changing aspirations and standards already alluded to. If housing associations have no competitors apart from others from their own sector, and possibly private sector bodies developing with a form of social housing grant, how is it possible to achieve the best results through comparison and contrast?

If a housing association, or even an RSL consortium, effectively holds a monopoly in social housing development, surely this is no incentive to drive standards up, or even to keep them at the same standards, unless the Housing Corporation comes over all heavy-handed with inspection and regulation on a site-by-site basis, which is highly unlikely, or the local authority building control department inspections reveal grossly substandard work.

The argument that housing association development is pound-for-pound more cost-effective than subsidising loans raised by councils to build, due to the fact that SHG rates are lower than 100% and frequently approach 50% of approved development costs is true only because of the government's policy of forcing associations to borrow the amount not financed through grant privately, and to limit grant to less than 100% of TCIs. Each proposition follows from the other, and is therefore only a fancy way of stating the present policy twice. If the government decides to limit grant to less than 100% of development costs, clearly the costs to the exchequer will be less than if the grant were 100%, or if loans were the only way of financing council housing development, and debts were subsidised 100%, taking the cost of subsidy over the entire term at net present value.

If the government allowed councils to borrow privately to develop, and insisted that they finance this borrowing through rental income, or financed a portion of the development through the application of own or nationally recycled RTB and land receipts, and only subsidised the residual public loan debt, they would be in the same position in accounting terms as housing associations, and it would be no more 'cost effective' in terms of public finance input for housing associations to build than for councils.

An example may help clarify the position.

Let us suppose that the cost of developing a house is £100 000. Suppose that the Total Cost Indicator for the dwelling is also £100 000, and that the grant rate is 50%. The housing association receives £50 000 SHG, borrows the rest, and has to meet the loan servicing costs from rental income. Leaving aside for the moment the question of where the money comes from to pay rents, given that there is often high housing benefit dependency in the RSL sector, and what the rents could be if grant rates were higher, the public finance commitment to the property in capital terms is £50 000.

Let us now suppose that a council develops a similar property at the same cost, and that it has no access to recycled or its own capital receipts, and has to borrow the entire sum from the Public Works Loan Board over 60 years. It will have to borrow £100 000 to back the development, which appears as a hit on the public accounts, the capital being accounted in the year the capital is applied. In simplistic terms, the hit on public finance is greater if the council develops than if the housing association does so, to the tune of 50% more. The rational macro-economic approach would be to go for the first method, given that this is the way the system is loaded.

If the local authority were to be allowed to apply capital receipts to 50% of the development costs, either from its own usable portion or from the national pool, the hit would be limited to 50% of development costs. The reason for this is that the capital receipt arose through a private purchase, funded by a commercial lending institution, and there is therefore no reason to account it as public money, as it was not borrowed in the first place by the Exchequer, or financed through taxation. If it is so accounted, it is an unjustifiable and arbitrary categorisation. It is true that to use capital receipts in this way would incur an 'opportunity cost' – roughly the value of what else could have been done had the sum not been applied in this way – but there are opportunity costs of the same magnitude associated with the decision to allocate part of the public resource to SHG rather than, say, to going for a hospital modernisation programme. The same would be true if councils were allocated grant in the same way as housing associations. The decision not to do so has no basis in rational macro-economics, but is rooted in political dogma. The other odd thing about this argument is that it relies upon the largely discredited dogma of monetarism – the idea already discussed that money is like any other commodity. If supply grows faster than demand, then its value will fall, which prompts inflation and destabilises the national economy internally and in relation to international trade relations. The argument runs that the government therefore has a duty to control the money supply, and in the absence of will to do so by credit controls or direct interest-rate intervention, can do so most effectively by keeping a rein on public expenditure and letting the private sector take more of the strain. One way to do so is to finance housing association development by applying grant at less than 100% (or nearer 50%) of approved development costs rather than by letting councils borrow from public resources to develop. Does this mean that less money has actually been spent – has been used – has been put into circulation – in respect of development? The answer is no: exactly the same amount of money has been spent, and put into circulation through paying for the development process. The decision to part-finance development through grant is therefore neutral in its supposed effect on inflation in monetarist terms.

Neither the Golden Rule or Sustainable Investment Rule would be breached by enabling councils to develop again: central borrowing associated with this would be to re-invest and not to finance current expenditure; and national borrowing could be kept below the relevant percentage of GDP by ensuring the recycling of capital housing and land receipts to keep borrowing within limits.

The argument that housing association development is 'naturally' a better means of using public money than council housing development on the basis of differential levels of public support only works because of the way the game has been set up, and has little or no external or objective justification.

The argument that common housing registers, allocations policies and regional choice-based lettings lessen the need to have more than one type of social housing developer is also questionable. It is true that, in the past, councils and associations let to their own stock, although councils have been nominating and referring to associations for very many years – in some cases, to 75% of RSL family lets and 50% of other types (exemplified by the *Partners in Meeting Housing Need* nominations arrangements between London RSLs and boroughs). But the existence of common lettings arrangements does not imply one supplier, or no need to have several types. The argument is sometimes advanced on the basis of economies of scale, and the costs of setting up council development departments again. Unfortunately, there are also diseconomies of scale, attained after a critical threshold has been reached in an organisation whereby it can no longer co-ordinate its various branches efficiently, where sheer size makes supervision of its functions more and more difficult, and where the cost of administration may unduly reduce the amount of resources which would otherwise be available for innovation and re-investment.

The argument that common lettings policies are spreading, implying that there is no longer a need for the letting body to build its own stock for its 'own' customer pool, does not point to any particular form of provider. The fact that housing associations are the only mass producer of social housing, either directly or through private building firms, does not mean that they are the only or 'natural' choice of provider. There is no link whatever between who provides the properties and who arranges for them to be peopled. The argument could equally be used to justify the abolition of housing associations, and to give the responsibility of building and allocating to one agency, say a decentralised or regionalised government department. There might even be efficiencies generated through this approach.

4. *Councils have built up a huge backlog of disrepair, admittedly largely through a consistent and long-term lack of central commitment to fund their activities, and restrictions on their ability to use their own financial reserves. It is better that they are left to concentrate on improving what they have, rather than to return to their pre-1990s role of developing houses for rent. Encouraging them to build new homes would dilute their efforts in this respect, would lessen the prospect of their meeting decent homes standards even where they have a prospect of doing so, and this risk is not an acceptable one.*

This argument ignores the fact that councils have, almost since inception, been multi-purpose vehicles, capable of fulfilling a variety of roles, ranging from the provision of education, refuse services, leisure and recreation, large capital projects such as road and bridge-building, through to the management of million and in some cases billion-pound budgets, all within the control of democratically elected boards and under central regulation and auditing. They are therefore capable of managing development as well as renovation programmes. The only constraint is the amount of cash they have to do it, rather than ability, or probability of success.

The privatisation of development?

It might be argued that the above is an over-robust defence of municipal housing born out of an ideological preference for non-commercial solutions to housing provision, and a bias against voluntary sector provision. However, all that has been argued is that the presumption against municipal development is questionable. Housing associations could take on the entire burden of social housing development, although, as has been argued, there is no good or overriding reason why they should, and a mixed economy in these matters might be preferable from a best value standpoint.

Could the private sector provide for housing need as or even more efficiently and cost-effectively than traditional social housing bodies, with minimal state support, obviating the need for councils and housing associations to get involved at other than the strategic or enabling level?

Approved development programme (ADP) resources have been advanced to private sector bodies, or housing investment trust vehicles since 2004. This is partly on the basis of claims by some private sector developers that they can produce affordable rented homes at rents equivalent to those charged by associations, or low-cost homes at prices comparable to the shared ownership or other discounted homes produced by associations with looser or in some cases no grant commitment; and partly driven by the continued desire to reduce the size of the public sector borrowing requirement.

It may seem odd to think that the private sector can produce affordable homes in any meaningful sense of the words more cost-effectively and to the same standard as a housing association in receipt of SHG. For one thing, commercial developers are for-profit organisations and housing associations are not-for-profit. Therefore, assuming that development costs are the same for each organisational type, it follows that a private developer would have to sell its equivalent property at a higher price than a low-cost home ownership association to satisfy its commitment to shareholders, as well as to meet financing costs and produce a surplus for re-investment, both of which might be true for a not-for-profit organisation.

In order to make a satisfactory return, and still to produce a unit of affordable accommodation for sale, it would have to make savings in its unit construction costs or other inputs, such as on the land price. True, savings on unit costs can be secured by bulk-purchase of materials rather than using the cheapest available components, and some commercial developers, for example, Bovis and Barratts, are extremely large and can command considerable economies of scale in negotiation of discount on bulk, but then so are some housing associations operating singly or in consortia. Compromising on the quality of components in order to hold down development costs and guarantee the shareholders' dividend would impose unwarranted ongoing costs on the occupier if an owner, or on the company itself in terms of replacement costs of components with a shorter useful life than those which were initially dearer, in the event that it leased or rented the units. It would also result in not providing like-for-like in qualitative terms (equivalent to reducing TCI's for housing association development and therefore reducing the quantum rather than the percentage of SHG). Either way, it is extremely difficult to see how private sector development could provide affordable housing for rent or for sale more cost-effectively – or at lower cost – than the RSL sector, due to the profit imperative.

The land-cost factor is an interesting one in this argument. Between 20 and 60% of the cost of development is down to land acquisition costs, depending on where the land is, and the level of competition of it. If one form of organisation can negotiate keener land prices than another, then it might be able to produce a product of equivalent quality genuinely more cheaply, even taking account of required profits, in which case why not give grants to such bodies? Again, it is very hard to understand why private sector developers might be in a better position than social developers in this respect. Unless they already own the land, developers have to acquire it at market price. Again, their sheer buying power might enable them to get it at a discount, but this also applies to housing association consortia. Even if they own the land, unless they are a charitable institution, they will seek to recover the value of the land used in development in the sale price, as land is an asset with value. Why would a profit-driven private sector company sacrifice the value of an asset to produce sub-market housing, when it could develop market housing at a higher price by reflecting the value of the land in the sale price?

The argument is sometimes advanced – and this is supposed to be a killer line – that private developers are inherently more efficient than social sector developers, because private enterprise is generally more efficient than the not-for-profit sector. Even if this were true, it is important to remember that housing associations almost never build homes themselves through direct labour organisations. They contract with private developers to do so, so any 'savings' through efficiency are passed on to the client in any case. Nor can the argument that, since the developer sells the properties to a housing association at a profit, and the housing association adds a surplus on to the eventual low-cost home-owner, or reflects any surplus in rentals, it would make for cheaper low-cost home ownership if the developer sold the property direct to the low-cost owner or rented it direct to the intending tenant, since housing associations are not-for-profit organisations and therefore do not add surpluses on in this way.

Enough has been said to cast doubt upon the ability of the for-profit sector to produce affordable housing of equivalent quality with or without grant more cost-effectively than housing associations. Time will tell whether, in the medium to long term, the experiment involving grants to the for-profit sector will actually produce low cost homes for sale or for rent more cost effectively or at the same or even lower costs to the ultimate consumer than can be achieved by the housing association sector.

The importance of sustainability

There has been much talk of sustainable development over the last decade, but little attempt to explain what it actually means: and yet it is a vital ingredient of any social housing development. Sustainability is a concept which can be exemplified as follows, and categorised into environmental, economic and social sustainability:

Environmental sustainability

- Sustainable building materials, i.e. materials which will not unduly deplete the planet's resources, which means the maximum use of renewable resources in building, such as softwood.
- Sustainable energy sources, i.e. use of appliances which run on renewable resources such as wind, water and solar power, or use of electricity generated by hydro-electric or wave power rather than fossil fuels, both to conserve non-renewable resources for the sparing use of future generations, and to help minimise global warming by not increasing the thickness of the greenhouse gas layer.
- Use of heating appliances which are energy-efficient, again to minimise the use of non-renewable fossil fuels and to minimise the production of greenhouse gases, to ensure the sustainability of life on earth in the medium to long term.
- Sustainable green or 'soft' landscaping: maximise the use of trees and hedges on sites in order to maximise oxygen production, to counter-balance the increasing use of oxygen by the world's growing population and to help neutralise the effect of increasing pollutants and greenhouse gases which inevitably contribute to global warming. The presence of trees, especially in river basin areas, can also lessen the chance of flooding, as relatively large

and complex vegetation slows down the passage of water in the hydrological cycle, moving from precipitation through to infiltration into the soil and translation into runoff. Evapotranspiration also returns water vapour to the air, thus increasing the time delay between condensation and precipitation, and therefore lessening the chance of flooding. Tree roots also suck up vast quantities of water. Placed too close to dwellings, trees can thereby cause shrinkage, especially in clay soils, through drying and can lead to foundational subsidence, but planted at a reasonable distance, they can soak up excess soil water, and lessen the chance of catastrophic flooding.

- Locational sustainability. Not building on unprotected floodplains, or those which are unprotectable in the medium to long term, due to rising sea levels as a consequence of climate change, for example estuarine areas, unless one wishes to go along the Thai model of stilthouse building, the Venetian model of building on sunken pile foundations, the Netherlands model of reliance on substantial sea-dams protecting polders and low lying areas, the floating town model of enlarged houseboats and other floating structures with connecting ferries ('Waterworld'), which would rise with sea levels, or even the Captain Nemo or Atlantis model of underwater living. These alternatives may have aesthetic or lifestyle attractions, especially for fishermen and lovers of fast boats and ocean views, but it would be a shame if these alternative futures were thrust on unwitting residents by accident.

Economic sustainability – the property itself

- Building houses and flats which are economical from the point of ongoing maintenance and repairs, so that landlords do not have to foot the costs of excessive repair bills and pass them on to tenants, and tenants and owners do not have to deal with the cost and inconvenience of components which fail before they should.
- Ensuring that heating and other appliances supplied are relatively economical to use, in terms of fuel consumption

and duration of use. This may be achieved partly through the use of double glazing and thermal insulation in cavities and roof voids, and the use of wet radiator systems with thermostat controls which vary the temperature automatically with changes in the outside weather. Water-using appliances should also be economical and waste free, especially where use is metered and paid directly on a pro-rata basis by the occupant, or recharged on a usage volume basis.

Economic sustainability – ensuring that dwellings are well located with respect to employment

- Employment sustainability. Dwellings should be built within reach of work places, preferably close to areas of industrial or other commercial vitality and expansion, to ensure that occupants will be able to obtain or maintain paid employment, without having to spend excessive sums on transportation.
- Housing costs sustainability. Every effort should be made to ensure that rent levels are as low as possible, consistent with the quality of the dwellings, to ensure that occupants can afford to pay the rent without recourse to housing benefit if in lower-paid work, and in relation to the household expenditure requirements associated with the type of households consistent with full occupancy of the dwelling, and in relation to the dwelling mix.
- Expansion capability: can the scheme be expanded if there is economic growth in the area and therefore greater in-migratory demand?

Social sustainability – to promote social harmony and avoid social dysfunctionality

- Age-related sustainability. Schemes should be designed to ensure that the child density is likely not to create a noise and behavioural nuisance to others in the community. This implies not only looking carefully at the number and age of children likely to occupy the dwellings, and crude density considerations in terms of bedsize occupancy, but also designing in children's play areas which can at once be supervised and draw noise and behavioural disturbance away from dwellings designed for single people, or elderly people who might not appreciate the presence of large numbers of children.

- Demographic sustainability. There is a case for ensuring that schemes are designed with a mix of house and flat types suitable for different sorts of household, to ensure demographically balanced communities, although this is sometimes advanced as unstudied dogma, raising the question as to whether there is anything inherently good about a mixed demographic community or housing scheme, especially where behaviours may be incompatible. This is an area where resident attitude surveys in existing developments can make a big difference to scheme layout and dwelling mix considerations.
- Tenure-mix sustainability. This is a complex question. It is frequently assumed that a mixed-tenure community is superior to a single-tenure one. The argument runs that single-tenure communities tend to produce ghettoisation and snobbery. Council estates, if dominated by tenants, may have a bad reputation locally because of the low esteem in which council tenants may be held by owner-occupiers who may unjustifiably perceive them as 'problem families', and some residents of such estates may accept the label and conform with the imposed stereotype, giving rise to real social problems in the area. If schemes are designed as mixed-tenure in the first place, with owner-occupiers, low and full cost, living next door to social renters, and with a degree of mobility between each tenure – usually from rented to owner-occupied, and where the dwellings are of equivalent quality, then the stigmatisation and polarisation described above would be less likely to happen, with fewer attendant social problems arising. The argument has also been advanced that owner-occupiers might inspire tenants to look after their gardens and look to their children's behaviour by superior example, or at least exert pressure on unruly tenants to conform to 'decent' behaviour standards, if only in a desire to maintain their property prices. Everyone

would gain from this – tenants would self-improve through good examples, owner-occupiers would maintain the value of their investments, there would be less anti-social behaviour, repairs and maintenance bills to communal areas and individual dwellings would be lower, and crime rates might also be lower as the alienation of tenants which might cause criminal 'protest' in the form of burglary and vandalism would be less. A veritable garden of Eden. The theories are all untested and probably lack substance, but who knows? There may be a correlation between mixed tenure and social harmony, if anyone has actually studied the matter.

- Lifecycle sustainability. Given the inadequate rate of production of new social and otherwise affordable homes in relation to housing need, and the low probability that there will be a massive expansion in the social housing sector, despite the size of affordable developments in the growth areas, it is likely that households in the high-demand areas of the South-East will have to stay where they are for much longer than hitherto, due to the lack of transfer and other mobility opportunity. This implies the need to ensure that dwellings are built to 'lifetime' standards, to ensure that they can be used throughout the lifecycle, without forcing moves on occupants. 'Lifetime homes' were developed initially in the Netherlands in the 1970s, a country with real land-supply problems, and have become standard for many RSL developments. The idea is that the same dwelling can, through minimal conversion and adaptation, provide extendable accommodation for growing families, possibly be convertible to smaller units when children leave home, and ultimately cater for the needs of elderly and frail tenants, through conversion and the provision of wheelchair-accessible door sets, ramps and level access. On the face of it, this strategy seems eminently sensible: any increased development costs over 'standard' demographic and household-type-specific dwellings may be offset in the longer term by eliminating or reducing void period on move-out and re-selection.
- Security sustainability. The scheme should be designed to ensure that crime is 'designed out' as far as possible,

and that therefore people will want to stay there, to ensure the long-term viability of the area. Police advice through 'Secured by Design' type initiatives should be factored in. This means avoiding blind alleys, unsupervised areas, car parking away from dwellings, underpasses and subways, paths bordered by fences or high hedges, and the installation of monitored CCTV, ensuring that there are not too many escape-routes from schemes, and definitely no walkways in the air, along with appropriate staffing, such as concierges in flats and possibly neighbourhood wardens working in close liaison with the police. Alice Coleman's *Utopia on Trial* (1986) is still worth reading as a guide to design-led solutions to crime and anti-social behaviour, although the role of design as opposed, say, to the involvement and management strategies advocated by Anne Power and PEP in the 1980s and early 1990s (PEP, 1986) which have in many cases produced more durable results.

- Avoiding 'sink' estates of benefit-dependent or unemployed households. This is a tough one, because inevitably the new customers of social housing organisations are likely to be those who cannot access other forms of tenure, specifically owner-occupation and to a lesser extent private renting due to wealth and income disadvantage, and who may otherwise have been homeless or forced to live in downmarket, cramped and otherwise unsuitable housing. Concentrations of such households may contribute to social problems on estates, borne out of perceptions of hopelessness perhaps, although much of the evidence for this is allegorical rather than real. Studies have shown that labelling by outsiders materially contributes to the reputation of areas as much and even possibly even more than the actuality, and that, far from constituting a socially deviant underclass, societal norms are often as elsewhere. Care must therefore be taken in advancing this argument, but adherents might lean towards the idea that a sustainable community equals one where tenure and demographic mix is designed in.
- Infrastructural sustainability. It is not enough to build vast expanses of dwellings without taking into consideration the 'liveability' factor, that is, what is it like to live

in such an area, will people want to stay, will they want to live there in the first place unless forced to do so? The question of proximity to economic opportunity has already been raised, but issues of proximity to transportation opportunity to workplace, leisure facilities, educational establishments, health and hospital provision, retail and services, and so forth, need to be considered carefully in the creation of truly sustainable communities.

At root, sustainability is all about considering the needs and aspirations of people first, and designing accordingly, rather than trying to shoe-horn people's wants and needs into convenient boxes and settlements which delight the planner and economist, but which bear no resemblance to where they would aspire to live, or actually choose to reside.

Sustainability is, then, a multi-faceted concept, and it is important to factor in all its aspects when undertaking social housing development.

The importance of planning

Residential development requires planning permission, which is related to the aims of the local authority's development plan. This area has been subject to change in recent years, and so it would make very little sense to go into fine detail, as the system will probably be unrecognisable well within the currency of this book. The Town and Country Planning Acts which provide a framework for the production of strategic and operational plans are well out of date, and can be expected to be subject to considerable re-drafting. Circulars and guidance such as Planning Policy Guidance Note 3 (PPG3) – *Land for Housing*, give valuable advice to local authorities, and are revised from time to time, principally in response to enable more affordable housing through the planning process in the context of increasing demand in growth areas.

There are two key aspects to the planning process which affect the design and layout of schemes, or even if it will take place at all. The first is the development plan itself,

London, Section 106 can in some cases be the only significant source of development land in an area. Where the cost of the land is very high, and RSLs struggle to meet their development costs within TCIs, it can be a godsend, even though Section 106 developments do not attract SHG. The advantage is often purely that access to land and housing is given which would not otherwise be.

It is important to note that planners, when negotiating Section 106 agreements and in fact when considering the grant of planning permission at all, cannot insist that the gain be taken in the form of social housing (if by this is meant housing available through housing associations and councils at below-market rents to meet housing need rather than effective demand), but can only refer to affordable housing, which is tenure-neutral, and covers lower-cost housing for sale as well as for rent. This is because the planning system is principally concerned with the exterior and interior layout of dwellings and the built form of schemes, encompassing design, road and path layout, green areas, provision of infrastructure, hard and soft landscaping, etc. The nature of tenure of developments is a preserve of the developer and housing strategist.

Planning, then, is an essential aspect of the development process, and debates rage over who should have the responsibility to devise strategic plans and operate development control. It seems reasonable that strategic plans should cover more than single districts, as well as take account of neighbouring structure plans, because the impact of developments, especially those of significant size such as major housing developments and out-of-town shopping centres often have a regional impact in terms of traffic generation, connection to markets, employment requirements nd social infrastructural needs. If it is desired to encourage igration or daily travel from a relatively less economically ealthy area with structural unemployment or underemyment due to the decline of a major industry to one ich is growing economically, then it might make sense lan for this on a regional basis, to ensure that the decg area does not suffer unduly and is linked into an area prospects by adequate transport infrastructure.

and the second is development control, which takes place within the development planning context. Development planning involves local authorities taking stock of the housing, employment, leisure and recreation and social infrastructural needs of an area on a periodic basis, and translating these into land use zones and development standards on the ground. Outside London, the first part of this high-level strategic planning, helping to define where given types of growth will be permitted and resisted, is a county council function, outside the unitary councils, with structure plans produced every ten years and revised every five. The structure plan has the disadvantage of having to cover a significant period of time, during which pressures on development may change, even over the review period. Structure plans are broad brush documents, which have to take into account both of those neighbouring counties, and the view of the new directly elected regional assemblies, as stated in their Regional Strategy Statements, which may soon (at the time of writing) acquire strategic planning powers. There must also be consultation with district councils which lie within the counties, both at an officer and member level, and with residents, typically through questionnaires and road shows.

District authorities produce detailed local development plans, which indicate precisely where certain forms of development, for example, industrial, residential, leisure, retail and office, will be permitted and prohibited. The key part of the document is the land use zone map, which defines the spatial location and extent of land use classes. The classes themselves are laid down by legislation, although there is considerable latitude on their application. Attached to the detailed justification of the zoning is a development standards document, which indicates the permitted density, height, design and other aspects of schemes subject to planning permission, which forms the basis for the council's development control function.

Local plans are typically produced every five years, but subject to regular review in the light of changing circumstances, such as demographic changes attendant on changing migration patterns, legislational changes, and shifts in the economic fortunes of the area.

Unitary councils such as the London boroughs, which currently combine all the functions of district and county councils, produce both strategic and operational plans in the form of unitary development plans.

The process of obtaining planning permission can be a very lengthy one. It is necessary to submit an outline plan, in order to receive consent in principle, before producing a detailed proposal for scrutiny. There is always the danger that the need for that particular development will have ceased or been modified by the time that planning permission is granted. There are, admittedly, timetables for the completion of the various stages, and for local authority responses, but it can take up to two years to get a proposal through the system.

Local authorities can make three basic planning decisions: the grant of planning permission – either detailed or outline, conditional planning approval, and rejection, although the latter carries with it a right to appeal. Major schemes may be subject to a planning inquiry, on the basis of size and impact on the existing developments and communities. Planning consultancy is big business, partly due to the size of the prize if planning permission is granted, and also due to its complexity.

Planning gain

Section 106 of the *Town and Country Planning Act 1990* is of key importance to the enabling of affordable housing on the back of commercial development. Like powers have been around since 1971 when the concept was introduced by the Town and Country Planning Act of that date.

It has already been said that the development plan forms the basis for consideration of grant of planning permission, and that developers who wish to create a scheme outside the envelope provided by the local plan can in most cases expect refusal, or a long and expensive process of trying to get a version through under the appeals process, with no guarantee of ultimate success.

However, Section 106 provides that planning authorities can negotiate with a developer, a process initiated either

by the applicant on the basis of an outline or worked-through proposal, or authority, to secure some community gain from varying the detail but not the spirit of the constraints of the development plan. For example, the case where a residential developer proposes to build at slightly higher density than envisaged in the local plan, or proposes to build houses or flats on a site designated by the authority as for commercial or retail use. The proposal must not be totally at variance with the constraints of the development plan, otherwise there would be no point in having it, but it is surely reasonable that there should be some latitude for variation, especially given that economic conditions may change, and therefore (perhaps) the need to retain a site as commercial may be obviated.

The idea is that the developer promises to share the development profits with the local community as the price for the council permitting development which might have been marginal, or refused normally. This sharing can be in the form of houses and/or land to a third party, for example a housing association, at discounted or no cost, or a cash sum to the authority – a form of 'planning tariff' which i
becoming ever more popular in lieu of other advantag
passed across on-site by developers. Where the gain is sta
as a planning obligation under Section 106, it becom
legal charge on the land. At the time of writing (2
there were plans to introduce a 'planning gain supple
a form of local tax levied on extra developmen
acquired as a result of such negotiations, instead of
itional requirement for land or housing on-site
Some developers might welcome this, as mar
the presence of affordable housing (especially
of social housing) might make it more difficu
market their product.

Section 106 is an extremely useful wa
affordable housing development into s
otherwise not contain it, and can obvi
for housing associations, where land, h
be acquired at nil cost, with the cost
being limited to management and m
shortage of residential developme
demand areas – for example in th

The need for rational regional planning is illustrated by the *Sustainable Communities Plan*, encompassing the Thames Gateway development, the expansion of settlements in the M11 corridor towards Stansted, expansion plans for Ashford, and the enhancement of Milton Keynes, all aimed at the management of economic growth and provision of housing and related infrastructure to meet the 2015 target for completion. In the past, the development of new towns around London, and those in the Midlands, North East of England, and central belt of Scotland all required a regional approach to planning, as each of these projects was an attempt to contain the pressures of growth evident in major urban-industrial areas, and had cross-local authority boundary implications, even at county level. The emergence of Regional Development Agencies in the 1990s to foster economic growth in the lagging regions of the UK also raised questions about the appropriate scale of infrastructural planning, and whether this could be left to even consortia of existing planning bodies at county level.

The successful economic regeneration of London's Docklands could not have occurred without the establishment of something like the London Docklands Development Corporation (LDDC), the urban development corporation set up there in the early 1980s by the Thatcher administration and under the direct control of the Secretary of State for the Environment, Michael Heseltine, MP.

It is perhaps unlikely that the boroughs in the area, even though they formed a consortium of convenience, could have delivered a unified scheme as rapidly as the LDDC, as each borough had significantly different planning policies and attitudes towards commercial expansion, although an umbrella planning development organisation with a board comprising borough representatives and other interested parties, but chaired independently, and with an executive with delegated powers, might have achieved similar results.

Advocates of regional planning bodies which sit above traditional democratic, locally elected and accountable institutions have to answer the question as to how the interests of those already in the area will be taken into account. If there is no direct democratic link between the decision-making or operational body and local people and interest groups, then many are bound to feel alienated by this process, and cannot be expected to be willing partners in the process of regeneration. The answer may lay in rigorous and sufficiently lengthy and detailed consultation processes, which expose plans to a variety of tiers of local representation, through the system of local government and out to the streets and homes of ordinary people in the form of questionnaires, focus groups, roadshows and 'planning for real' exercises. There must also be the possibility of local people influencing the end result, as they will be consumers of aspects of the product. It will affect their lives, and not always positively.

That is not to say that regional bodies with planning powers cannot be responsive to local concerns. Much depends on the degree of devolution of power from the centre, and the forward plans for these bodies. Many of these have now been disbanded, their job done, and power handed back to the local authorities from whom it was wrested, and sometimes the organisations set up by a central body have morphed into councils, as is the case for the new towns, which started life as projects of the centralised Commission for New Towns, which espoused decentralised management in each of the settlements, and controlled their affairs through localised boards. There is no reason why an enabling body should outlast the thing it has enabled, but is it has not engaged with the local institutions in a meaningful way, ownership of the final product may be taken on unwillingly, and not nurtured in the way it would have been had there been ownership from the outset.

The development process

The process of housing development comprises the following stages: establishing the need; examining development options; costing out the development and raising the necessary finance; planning considerations; site acquisition; design; construction and dealing with defects on completion, and, perhaps most importantly, learning from the experience so as to do even better next time.

The first four stages have already been discussed, with sustainability considerations covering the issues of density, dwelling-mix, layout, accommodation for cars, tenure mix and the use of appropriate materials which should be taken on board at the development options stage.

Land acquisition

Assuming that planning permission has been granted for a scheme, only then is it safe to acquire land. Admittedly, the practice of land-banking, that is, buying land with or without planning permission in the anticipation of developing it later, or perhaps to maintain an artificial shortage of development land in order to keep the price of the commodity and the product built on it relatively high, has been common practice in the last twenty years, especially in high development-demand areas, but it is a risky proposition in any case, and social housing organisations cannot generally afford to do this.

The sources of land are many and various.

It may be owned by a local authority or other public body, large private landowner or a number of smaller landowners. Who owns the land may well affect the price to the purchaser, as may the nature of the purchaser. A local authority may prefer to sell to a housing association at a discount in return for high nomination rights to the dwellings. Local authorities have to dispose of their assets for 'best consideration' and justify any discount, sometimes to the relevant Secretary of State, and nomination rights are held to be a payment in kind for land discount. The exception to this is the Right to Buy, where public assets have to be sold at a discount to sitting tenants, although there are limitations on the amount of discount available to protect the public interest. Local authority land sold at discounted price can also reduce the rate of Social Housing Grant available; and even land secured at no or low costs through Section 106 can reduce or even eliminate entitlement to SHG, albeit on a case-by-case basis subject to Housing Corporation discretion.

It may be owned by another public authority, for example the Department of Health and Primary Care Trusts, which have disposed of numerous hospital sites over the years to private and social developers. Sometimes, preference has been given to social housing developers, either individually or in consortia, or land has been sold on condition that some of it must be used for affordable or social housing development, or for the use of 'key workers' such as health workers or police personnel. Government departments have to get best consideration for their assets. It makes sense to utilise public sector land for the benefit of lower-paid but entail public service workers, so-called 'key workers', because they are needed everywhere, and this is often the only way to ensure affordable housing relatively close to workplaces.

One source of publicly owned land which can make a significant development contribution is that owned by the Ministry of Defence. The armed forces have been scaled down significantly in recent years, releasing land and housing formerly or still occupied by service quarters. Some of this land is at distance from other settlements, and may not be suitable for use for affordable housing development, because it is inconvenient for work travel or otherwise poorly linked to existing infrastructure, but other sites are relatively close to towns and cities, or close to major roads. In some cases, parcels of defence land have been sold at undervalue to housing associations and to the private sector, sometimes with conditions attached that at least a portion of the new development be used for affordable housing. Some of the housing could also provide spacious homes for lower-paid households, and there have been disposals to social housing organisations, especially where there is no security implications because bases have been relocated or decommissioned. The sale of public land was given fresh impetus by the 3-year Spending Review 2004 (see HM Treasury website for overall Review issues, and the ODPM website for housing and related targets), which indicated an acceleration in the disposal of publicly owned land and other assets from 2005 to 2007. Much of this land will, in the main, be sold by the government's agency, English Partnerships, to residential developers, including housing associations.

Land owned by private individuals or companies is rarely sold at less than market value, although prices vary locationally, often depending on where they are in relation to existing settlements or transport nodes. Much also depends upon the use class to which the land belongs. Agricultural land invariably sells for less – as land classified for agricultural use – than it would if there were permission for housing or commercial use. In countries which consistently produce an agricultural surplus in relation to demand, such as the UK or Holland, there has been a policy of 'set-aside' under the European Union Common Agricultural Policy (CAP) since the 1980s, and in these cases, agricultural land has been taken out of production, sometimes permanently, with farmers collecting subsidies for not growing crops on the land.

Under these circumstances, it makes sense to view such resources as potentially available for housing use, subject to conservation limitations, if any. Instead of paying subsidies to farmers to take land out of production and thereby incentivise keeping the land unused, or using it for paintballing, car racing or as nature reserves, it might make more sense in selected instances to pay a subsidy to allow the farmer to sell the land for affordable housing development at less than market value, with the grant making up the difference between agricultural market value and the sale price. It would cost the EU no more than the present policy, and release valuable development land to help resolve housing crises in many parts of the country and across Europe, and would also guarantee a capital return to the farmer. The problem of unaffordability in the countryside is in some areas just as pressing as urban unaffordability, and is a powerful force to instigate 'push' migration to towns and cities, where many rural occupants would far rather stay in the countryside, as long as there are jobs there. The revised CAP subsidy system outlined above would enable farmers to make a contribution to solving this problem, as well as compensating them for reducing the size of their landholdings.

Brownfield and greenfield sites

Greenfield development

Some land has never been developed for housing or otherwise, apart perhaps from agricultural use. Sites which occur in this context are known as 'greenfield' sites. Their development potential depends not only on their location with respect to neighbouring settlements and existing social and economic infrastructure, but also on conservation issues, which rural protection pressure groups such as the Council for the Protection of Rural England (CPRE) frequently draw attention to in challenging residential development growth proposals.

In areas of relatively high population density, especially around towns and cities, there are good reasons for resisting greenfield developments, which go beyond the NIMBY (not in my back yard) pressures exerted by existing residents who have moved to the countryside to escape the pollution and noise and other stresses of urban or suburban life.

These include the reasonable desire not to let settlements coalesce unduly, on grounds of transport management and retaining the separate identity of places, as well as the costs of providing social and economic infrastructure to ensure that such growth areas are sustainable. The prohibition or restriction of development in corridor areas around cities to provide leisure and recreation opportunity to existing urban settlements, and to act as a restraint of further uncontrolled growth, lay behind the greenbelt policies pursued from 1947 onwards in respect of London and other major urban centres. The effect of greenbelt policy has been to preserve the countryside as a valuable amenity, but also to increase the price of land and new housing there, due to the premium that shortage commands under high-demand circumstances. Parcels of greenbelt land are occasionally released from residential development restrictions by variations to planning by-laws and regulations, and when this happens, the resultant development is generally expensive due to high land values resulting from planning-imposed scarcity, unless it is publicly owned land and a deal has been struck with an affordable housing provider.

The development advantage of greenfield sites is that the land is generally uncontaminated, and free of existing structures: therefore decontamination and demolition costs are negligible, and developments can generally be realised more rapidly than otherwise. It also makes layout and design much easier. The disadvantages include the problem of supplying social and economic infrastructure – new schools, hospitals, libraries, roads, shops etc., unless the site is adjacent to an existing settlement, although the facilities of that place may well have to be enlarged to cope with increased need and effective demand. There is also an opportunity cost, as there is with all other development: what else could the site have been used for? Housing development rules out an easy return to agricultural use, or use for early industrial or other commercial development, for which there may be a renewed need in years to come, and it is important to consider the alternative uses for sites when granting planning consent, or changing land use classes.

Additionally, there is an environmental opportunity cost, as well as the importance to human wellbeing of working with rather than against environmental and ecological factors. Greenfield sites are already occupied – by birds, plants and animals, which form part of a food web. By destroying natural habitats in one area, there may be knock-on effects on neighbouring ecosystems, which may lead to an escalation in the number of insect and other vermin pests through the destruction of their natural predators' habitats. There may also be a disruption to the water cycle, which may lead to increased flooding risk by disrupting the infiltration and percolation of water into subsoils through concreting and tarmac layers, which are generally impermeable.

The removal of surface water storage hollows and diversion of water courses may also lead to increased flood-risk, or indeed to the drying of streams, depending on the water requirements of, and extraction arrangements for, new settlements. Removal of trees and shrubs will also reduce evapo-transpiration rates, the return of water vapour to the atmosphere through the combined effects of evaporation and the release of water vapour from the stomata of plants, which increases the rate at which precipitation flows through the cycle into the ground, again increasing the problem of runoff control, and necessitating extra drainage and flood protection measures which can be very expensive, and in some cases unsightly.

Localised changes in the colour of the Earth's surface, by substituting dark tarmac or very light concrete, or reflective surfaces for green, will also have an effect on local microclimate. Heat is stored in dark tarmac or asphalt surfaces by day during the summer, and released at night, rather like a storage heater, to a greater extent than is the case for vegetation, and this can create rising thermals. In this case, if relatively moist air moves over these surfaces at night, it will be moved upwards rapidly by these heat currents, causing convection rainfall which can be heavy and sudden, and can test the capacity of drainage systems, leading to flooding and infrastructural disruption. Similarly, very large settlements create a 'heat island effect', where air temperatures at night are often several degrees higher in the centre of the settlement at night than at the edge. This may be very convenient for barbecues and pavement café society, but the effect of this can be the retention of moist pollution-laden air – smog – in the 'heat island' dome, where layers of comparatively warm air are trapped under colder, denser air further up. Normally, such pollution would disperse, and therefore such man-made heat islands can be prejudicial to health, causing respiratory diseases. This necessitates careful thought about urban densities, and the provision of green spaces, trees and water masses, in new settlements of significant size.

For all these reasons, greenfield development needs to be planned and executed sensitively.

Brownfield development

Brownfield development occurs when land which has already been used for industrial, residential or commercial functions becomes vacant once more through demolition or because there is no longer a need or demand for that use, and is redeveloped. During the 1990s, successive administrations stressed the importance of using brownfield land for re-development wherever possible so as not to make excessive demands on the countryside. This makes a lot of sense in principle. After all, brownfield sites are often close to existing social and employment infrastructure, and next to existing communities, which increases the chance of subsequent housing developments being socially and economically sustainable without provision of new employment. It also removes the eyesore of ugly, derelict buildings which have had their day, and can revitalise surrounding communities by providing expansion opportunities. Buildings on brownfield sites may also in some cases be converted to another use – witness the conversion of warehouses beside the Thames in London's Docklands in the 1980s and early 1990s, and the successful conversions in Swansea, Cardiff, Bristol, Liverpool, Manchester and Salford. The use of brownfield sites in this way is a major contributor to urban regeneration, and has revitalised many hitherto worn-out inner city areas.

The disadvantages of brownfield sites as redevelopment opportunities include the costs of demolition, and re-instatement of the land. Building on contaminated land can lead to health risks later, and the presence of decaying industrial waste beneath foundations can cause ground instability, and massive latent defect remediation bills later. The build-up of methane gas from effluent and dumped rubbish sealed under the concrete of new developments can lead to potentially explosive situations, a potent thought for those walking home south-westward from the wine bars and cafés of Fulham Road and Chelsea after a night out in one of London's more fashionable districts, for Fulham is built on the site of a pre-Victorian rubbish dump. Much stress is therefore laid on ground testing, searching for historical records of contamination, and decontamination in planning and building regulations. Dealing with contaminated land can add significantly to development costs, and make a scheme unviable, although the desirability of bringing land back into use and revitalising decaying areas may outweigh the costs, and justify government grants to assist the process.

In summary, the re-use of surplus used land and buildings makes environmental sense, if it can be realised effectively and economically, as it obviates the need for urban sprawl and ecological disruption. It also makes sense from the viewpoint of community sustainability.

Land, then, is of major importance to development, and not just because it is very difficult to build in mid-air and at sea, but because it can represent easily one half of total scheme development costs.

Design – layout and dwellings

Design happens before land acquisition, and can be the make or break factor in obtaining planning permission. The efficacy of design can certainly have a major impact of the lives of those who eventually live there, and those already resident in the area. Design can be looked at in terms of appearance and functionality. Appearance is a difficult one, because aesthetic tastes have changed so much over the years, and beauty is probably in the eye of the beholder, although what is acceptable in appearance terms at any one time can usefully be determined by customer surveys, especially amongst groups who might expect to be ultimate users, as well as by neighbouring residents, which is one reason why examination in public and consultation are held to be so important in the town and country planning process. Most planning authorities have design standards which address aesthetic as well as height, density and structural considerations. A golden rule in planning is to consider the visual impact that a proposed development will have on adjacent structures. Would you put a skyscraper next to a Georgian mansion? Well maybe – look at the skyline of London. Would you allow the development of a six-storey shopping centre next door to Buckingham Palace? Probably not, although there is no accounting for the ways that aesthetic tastes change.

As far as social housing is concerned, design means far more than aesthetics, although this factor should be considered. Just because social housing developers are catering for those who probably have no choice other than to have recourse to the sector, possibly excluding customers for low-cost home ownership, does not mean that the aim of attractive design should be thrown to the wind in the interest of cost-effectiveness, through packing as many dwellings in as small an area as possible by cramming houses together or building to thirty storeys to save on land costs.

Design means looking at density, materials used, appliances installed, and at internal space standards, to allow users to live there economically and stay there for a very long time throughout their life cycle if necessary.

Density

Density is a difficult issue, but rarely determines the success or failure of a scheme in isolation from other factors, except possibly at the extremes. It can be measured by the number of persons per area (e.g. per hectare), number of habitable rooms per area, or even by looking at the amount of space available to individuals against some 'liveability' or crowding standard per area. High-density developments can often be very successful, and low-density ones disastrous. A four-star hotel may be vast and developed at very high occupancy densities indeed, but as long as the rooms are spacious and well appointed, the cleaning and catering facilities are sufficient to the task, the services work and can take the strain of the use-level implied by the high occupancy of the building, there is sufficient staff of the right calibre and training, there are sufficient escape routes in case of fire, and the communal facilities are luxurious, there is no reason why the development should not succeed and provide an excellent albeit temporary life experience, especially if it is in Knightsbridge.

Similarly, a block of flats at very high dwelling densities, even a high tower block, need not necessarily be a bad thing, as long as the right level of services and management are provided, and the interiors of the homes are commissioned to high standards in relation to space standards and services compatible with modern expectations, there is easy access to exits, there is very good security provision, such as CCTV and a concierge, communal parts are regularly cleaned, there is access to leisure and recreation facilities and to green spaces, and the properties are allocated in a mature and sensitive way in a sensible mix of households appropriate to the accommodation. High-density living breaks down when services fail or management arrangements fail, and not because of high-density living as such.

Sometimes, a high-density development is the only way to make social housing work in a given area, if very high land costs cannot be avoided and there are few alternative sites available. This is an area where the views of existing and potential residents can usefully be sought in managing expectations and ownership of designs, and has been achieved in some contexts through 'planning for real' exercises, where residents indicate how they would approach design and development, after appropriate training.

Design matters

There has already been some reference to design matters in the section earlier in the chapter on sustainability. It is essential that homes are designed for living in. This may seems fairly obvious, but anyone who has ever lived in a house or flat – and that is most people – will know to their cost how inconvenient dwellings can be from the user point of view.

Dwellings can and have been adapted to meet the diverse requirements of modern life. Those designing today's homes need to ensure that they are flexible enough to enable people to stay in them for as long as they want or are forced to so by housing shortage, and this means asking people what sort of homes they need to meet their lifestyle needs and aspirations, The 1961 bible of social housing design, '*Homes for Today and Tomorrow*' (the Parker Morris Report), which arose out of the deliberations of the government-initiated committee of the same name, proceded from an analysis of lifestyles to design recommendations, but did not major on customer survey.

The best house designers are those who will have to use them, and architects and planners really need to do much more in respect of consumer consultation ensure that what they design is suitable for its eventual and long-term use, even to the point of designing in the possibility of conversion from one form of dwelling to another, subdivision or compartmentalisation, and extension. The 'liveability' agenda – relating not only to the layout of estates but also to the design of homes – has been very fashionable in recent years, but is another way of stating the obvious – that houses and flats must be designed as homes for use rather than as a monument to the eclectic ideas of some architect or another.

The principles of good design can be summarised under the familiar headings of who, what, when, where, how, and why?

Who (or what type of household) is the dwelling to be designed for? Have their needs been properly researched? Have focus groups been formed to discuss design? Has notice been taken of customer surveys of dwelling usage, and have future needs been adequately anticipated?

What are the constraints impinging on the development – financial, materials, land availability – and how best to reconcile these with meeting household requirements? What are the incomes of the occupants likely to be? Best to ensure that the dwellings are well-insulated and heated efficiently, so that fuel bills can be kept down as well as damage to the greenhouse layer.

When are the dwellings needed? Is there an urgent requirement – in which case, perhaps modern methods of construction – using pre-fabricated elements constructed in a factory and assembling them as panels on site – can speed up the delivery process, as long as quality and long-term viability is not compromised.

Where are the dwellings going to be built? It would make sense to build them in a style compatible with their surroundings and to observe any conservation requirements.

How are the dwellings going to be built? Traditional or non-traditional? Will the building method itself compromise or constrain the use to which the dwellings can be put, or shorten their life? How can we ensure that the building process is as efficient and effective as possible?

Why? Why build in this way: would another way be better? Why use these designs – do they really match lifestyle requirements? Have we contrasted and compared with other possible designs enough, and asked enough ultimate users for advice on what is required? Have we learned from past mistakes, or taken on board the fruits of good practice sufficiently?

In short, design is far too important to be left to architects and quantity surveyors. The true experts are those who will have to live in and manage the dwellings, and engagement with the ultimate user is the key to success in effective residential design and development.

Contract management

Contractual matters can only be dealt with at all effectively in a text dedicated to the matter, and are of interest only to commissioning clients. Most housing professionals get nowhere near a building contract, or if they do, periodically inspect the site with or without residents. Contractual matters cover the legal relationship between the client, contractor and sub-contractors, and the means of effective contract supervision. An outline only will be given here.

Essentially, clients can go for design-and-build contracts, or more traditional client-contractor building contracts. The former are very popular in the housing association world, and involve choosing a design or designs from pattern books provided by building firms, with minor modification, and letting the firm get on with it, paying for the completed product at the end of the process, and requiring remedy of defects at and after the end of the contract. These contracts have the advantage of simplicity, and supervision is left entirely to the contractor.

Alternatively, contractors can be hired under the terms of a client-supervised building contract, where the client has significant supervisory responsibilities, perhaps through an architect, quantity surveyor and, on site, through a clerk of works who is the 'eyes and ears' of the client. A common type of contract is the JCT in the appropriate edition – the modern versions are descendants of the 1981 edition – with or without quantities (essentially a list of priced-up building components and labour needs for the job). Usual contract law applies in seeking remedies and injunctions.

Resident involvement

There has already been mention of resident involvement in development, but it is absolutely crucial to the long-term success of the scheme. How and when should residents be involved? The answer is, at every stage of the development process, even if they will not directly benefit from the scheme itself – as expert advisors, backed by training, clear information and, where needed, the services of a competent independent resident advisor.

The importance of resident involvement can be identified at the following stages.

1. Planning the scheme

People will have to live in the properties constructed, and operate in the environs. What better justification for involving them at the very start of proceedings? Resident involvement can take the form of forming a focus group of residents drawn from the social housing landlord's tenants and leaseholders, to advise on layout and design matters. This may have to be supported by training, and assisted by the services of an independent resident advisor.

Residents can be taken round similar schemes, and given the opportunity to speak to residents there, to get an idea of the advantages and disadvantages of living in the property types under consideration and the planned layout, so they can constructively input into the planning and design process. They can also take part in planning for real exercises, where they can model alternative internal layouts and estate designs, and discuss the pros and cons, supported by relevant case studies, and expert advice. They can also be made aware of the running costs associated with alternative designs, in the light of impacts on rents and service charges, as well as the development timescale.

This period can also be used to acquaint them with the rudiments of contract supervision. It is likely that these residents will not themselves be the eventual occupiers of the dwellings, or even be living adjacent to them, and so it will be necessary to incentivise their input by making expenses and time payments available, just as money would be allocated to any consultant advising on these matters. A spin-off from this process can be the strengthening of existing participation arrangements in housing management, development and strategy, and may help the social landlord in the process of reviewing participation compacts. It will also help empower residents to take part in major repair, refurbishment and regeneration projects which may be applied to the homes they live in.

2. Managing the contract

There is nothing to stop social landlords from supplementing supervision by the architect, quantity surveyor and clerk of works by employing residents to visit the scheme from time to time, and helping in progress chasing contractors, as long as this is done in a structured and informed way, and in liaison with the contractors. An element of resident involvement can be written into the building contract arrangements, and residents can sit on the board which receives formal reports on progress, and which authorises sign-off of payments at start on site, completion and post-snagging final payment. Again, this can pay dividends when it comes to major works or refurbishment contracts on their own homes. Another spin-off for residents is that new or enhanced skills will be given, which could help them in the jobs market and take them off housing benefit where possible.

3. Post-occupation satisfaction survey

Residents are experts in the critical appraisal of newly completed homes, as they are the ultimate users of accommodation, so why not use their expertise in devising resident satisfaction questionnaires and leading focus groups formed of the new residents or occupants of converted or refurbished homes? It is likely that the questions they come up will be more relevant than those devised by professionals who probably do not live in the house types on offer, and the lessons learnt from the post-development period will be that much more useful when the organisation builds or refurbishes again.

Radical involvement through self-build.

The fullest expression of resident involvement in layout and design is through self-build projects. There are many examples of this, ranging from the construction of shanty town settlements in lesser economically developed countries, to the self-build schemes popularised by the Canadian architect Walter Segal in the late 1980s and 1990s. The latter schemes, a prime example of which is to be found in the London Borough of Lewisham in Forest Hill, South East London, often make use of prefabricated or modern methods of construction, with the use of timber framing and renewable materials, and residents' rents can be set at a lower level to take account of the value of labour input, or sold below market value in recognition of 'sweat equity' – the implied value of the work undertaken.

Conclusion – the future of development

It is likely that, given the case of housing stress in high-demand areas of the UK and elsewhere, and other pressures on land use as already identified, that hard decisions will have to be taken on density and space-standards in tomorrow's homes. Land is not infinitely available, and comes at a premium – and its price directly affects the level of rents and the cost of owner-occupation. In order realistically to accommodate households in high demand areas, short of trying to change the economic circumstances elsewhere in order to provide a counter magnet to in-migration or to divert demand elsewhere, it is likely that the imperative to build up rather than along, to find ways of fitting social and economic infrastructure in with residential development in ways which may have been unacceptable perhaps even twenty years ago, and to rethink the provision of communal areas and facilities, will create very real challenges to housing developers and designers. To effectively produce homes which are at once suitable to households throughout their lifecycle, and which are affordable, will be very hard, and so now is the time to think through how to involve residents and ultimate users in the very tough decisions on quality and quantity which lie ahead.

The problem is no less intense in areas of lower demand. The tough decisions to demolish perfectly good homes to reduce numbers to the level of projected demand, and to remodel neighbourhoods so that multi-tenure demand can be effectively catered for in a sustainable manner, have to be taken, and it is better that they were taken with the involvement and consent of those who live in these at-risk communities. Lateral thinking will be necessary to factor in social and economic infrastructure which will make such places attractive to live in, and the mistakes of the past will have to be taken on board.

Housing provision is not just about housing, and reaches into the arena of economic planning in a spatial context, to ensure that jobs are there to sustain communities, along with accessibility to shopping, transport, education, public services, leisure and recreation – in short all of the ingredients that everyone looks to in the neighbourhood in which they wish to live. Hopefully, the planners of tomorrow will take on board the mistakes of the past when planning the communities of the future.

References

Chartered Institute of Housing (1991). *Homes for the Future.* CIH/RIBA.

Coleman, A. (1986). *Utopia on Trial.* Pergamon.

Habinteg Housing Association (2004). Lifetime homes: 21st Century Living. www.lifetimehomes.org.uk

ODPM (2003). *Housing Statistics 2003.* HMSO.

Power, A. (1986). *The PEP Experience.* LSE.

Bibliography

Goodchild, B. and Syms, P. (2003). *Between Social Housing and the Market.* RICS.

Housing Corporation (2003). *A Home of My Own* (The Report of the Government's Low Cost Home Ownership Task Force).

Housing Corporation (2004). *Affordable Housing: Better by Good Design.* HC.

Housing Corporation/Countryside Agency/Country Land and Business Association (2004). *Affordable Rural Housing.* HC.

Minton, A. (2002). *Building Balanced Communities.* RICS.

RICS/Housing Corporation (2001). *Delivering affordable housing through the planning system.* HC.

Royal Institute of Chartered Surveyors (RICS) (2004). *Housing market renewal – making the Pathfinders succeed.* RICS.

Websites

1. Royal Institute of Chartered Surveyors: www.rics.org
2. Housing Corporation: www.housingcorp.gov.uk
3. www.lifetimehomes.org.uk
4. www.englishpartnerships.co.uk
5. www.odpm.gov.uk

5 Summary

The themes of this book have been the assessment of supply and demand for social housing, its management and maintenance, how it is financed, and how it is developed. This summary identifies the key points made.

Supply and demand

Social housing exists because not every household can afford to satisfy their housing requirements through the market, due to supply–demand imbalances and unfavourable earnings/house price ratios, and because society deems it just to provide housing for those who cannot compete effectively to secure their own home. The measurement of housing need, which is a subset of demand, varies, and the reason for its measurement is to ration social housing, which indicates that there is just not enough of it available, and that social housing is a scarce resource which has to be allocated on the basis not of demand but on the basis of prioritisation.

Social housing organisations supply properties for rent and for low-cost sale according to local and national indicators of housing need, but do not operate in a vacuum, and a robust and adequate house building programme to supply effective demand is complementary to providing on a non-market basis, as increasing the supply of housing in areas of high demand may decrease the price of the product and render it more affordable, reducing the level of housing need, and making the task of social housing provision that much less onerous.

Housing management

The style of housing management – by local authorities and housing associations – has changed vastly over the last hundred years, driven by the interlinked pressures of legislational and social change. There has been a move away from the paternalistic Octavia Hill-type management styles of Victorian times, to a customer-centred approach, and there has been a shift in the ownership of management towards residents themselves, as shown in the co-operative movement and later through Tenant Management Organisations, where this is desired. Housing management is a complex task, and is governed largely by the provisions of the Housing and Landlord and Tenant Acts which inform the conditions of Secure, Assured and Assured Shorthold tenancies. Housing maintenance is part of the management task, and landlords cannot expect tenants to respect them unless these obligations are carried out efficiently and effectively.

Central policy changes have forced social housing organisations, especially councils, to contract out their management services or to consider and test other ways of managing their resources, especially where they are unlikely to put their property into a decent condition by 2010. The result of policy shifts have been large-scale voluntary transfers to housing associations, specially made or pre-existing, entailing a change of ownership and management, shifting management across to tenants themselves under the Right to Manage, to special companies such as ALMOs, to regeneration bodies such as the now-defunct Housing Action Trusts, and to PFI operators, as well as to private sector companies in some cases. It is likely that this diversification will at least enable the effective testing of different strategies aimed at improving the level of management and maintenance services to the extent that central regulatory bodies expect, but the ultimate judge of effectiveness is the end-user, whose views and life-experiences should ultimately determine which strategies will eventually prevail.

Housing finance

Finance is at the heart of housing operations, and the cash-limiting macro-economic policies of successive governments have meant that local authorities have in general struggled to resource services and maintenance. As a result, the council housing stock has deteriorated in many areas, forcing a drive to achieve Decent Homes status by 2010. Adequate and monitored investment may well have obviated this measure. Housing Associations, supported by social housing grant, are the main suppliers of new social housing, but there are alternatives to this monoculture, and no reason to presume that councils cannot be as effective developers as their voluntary counterparts. The rules on capital and revenue finance are complex, but the areas are clearly interlinked, and only a root-and-branch reform will deliver the quantity of resources and their effective use, which is necessary to support the level of social housing construction and management improvement programmes that are required.

Housing development

Housing development programmes must be based on robust housing requirements studies if they are to be delivered to the correct level both in time and geographically, and must be interlinked with spatial economic and social trends if they are to produce sustainable communities. Planning is an essential prerequisite to housing development programmes to ensure that developments are located in the correct place with regard to existing settlements, are of a capacity likely to meet requirement, and are functionally and aesthetically appropriate. It also has a vital role in securing affordable housing on the back of private speculative development through Section 106 powers, and its role, through Planning Policy Guidance and other steerage, is of growing importance.

Social housing, at its best, is developed in response to customer needs and aspirations, and resident consultation at all stages of the process can assist in producing homes for today and tomorrow.

All in all, each of the key themes are interlinked, and an understanding of them, and the way they link together, is a prerequisite for intelligent and effective engagement in the social housing sector. The conscious linking of these themes is the key to sustainable progress in the sector, and their disaggregation means certain failure.

Glossary

Words or concepts are defined in the order they appear in chapters. Some definitions appear twice, to help the reader when reading each chapter in sequence.

Chapter 1

Rent restructuring Policy introduced in 2002 to try to relate social housing rents to market value, affordability and size in terms of bedroom-number. Objective was to complete the remodelling by 2010, and to 'harmonise' the rents of similar housing association and council-rented homes – that is, make them similar.

Effective demand Demand for a product or service which consumers are able to pay for, as opposed to the total requirement for that commodity or service.

Sector (employment) A way of dividing the economy up by type of industrial activity; e.g. primary (extractive, producing raw materials); secondary (manufacturing); tertiary (e.g. sales, administration and services); quaternary (industries servicing the former categories, e.g. banking, insurance); and quinary (information and communications technology), often included in the quaternary classification.

Green belt policy Policy introduced in 1947 to restrict development on and further out from the edge of urban areas to stop them growing to an unmanageable size, to provide countryside leisure and recreation opportunities for urban dwellers, and to conserve the countryside.

Housing Revenue Account (HRA) A housing management and maintenance account, showing current income and expenditure, which local authorities have had to keep by law since 1936.

Revenue Support Grant A payment from central government resources to support council expenditure, supplementing taxation levied by local authorities (council tax) and locally collected, but centrally redistributed, business rates income.

General Fund The local authority's corporate account into which central government subsidy, locally collected taxes and redistributed business rates, and some other charges, are paid, and which pays for local services and capital works. The HRA is a 'ringfenced' part of the General Fund – an account within an account which is used to fund housing landlord services, and which cannot be cross-subsidised from the General Fund, within which the HRA sits.

Housing benefit A locally administered income support benefit supplementing the income of rent payers in the private and social rented sectors who are judged unable to afford all or part of their rent. It is calculated on the basis of income, savings, household composition and rent level. See Chapter 3 for plans to 'reform' it.

Poor law A system of financial and related relief (for example, accommodation in 'workhouses') for households without the means to support themselves in defined localities (usually parishes), funded from taxes levied in the locality and administered by a board of local worthies. The level of support depended entirely on the views of the local councils, and their predecessors, who set it. It originated in the Middle Ages, was reformed in Victorian times, and all but abolished in the early twentieth century. After 1945, it was replaced by the modern system of welfare benefits, worked out centrally and administered locally.

Intentionality In the context of homelessness legislation, someone is intentionally homeless if they have done something, or failed to do something, which results in the loss of their home.

Priority categories In the context of homelessness legislation, a homeless person has a priority need for accommodation if they are pregnant, have dependent children, are aged or infirm, have lost their home through fire, flood or other natural disaster, or are 'vulnerable' for some reason, as defined in legislation, case law and government guidance.

Asylum seekers People who have arrived in a country from another country or state, claiming to being fleeing from persecution and/or danger or other peril, and who are thereby seeking 'refugee status'.

Registered social landlords (RSLs) Housing associations who are registered with the regulatory body set up by Parliament, the Housing Corporation. The term was established by the *1996 Housing Act.*

Office of the Deputy Prime Minister Department of State responsible for council housing funding and regulation. Works closely with the non-governmental organisation, the Housing Corporation, which regulates RSLs.

Quasi non governmental organisation (QUANGO) A regulatory and/or executive and/or funding body set up by and responsible to Parliament rather than directly to a Department of State, run by a selected board, which is voted public funds on a periodic basis. They have specific powers and duty, usually in relation to specific organisations, issues, or sectors. The Housing Corporation is an example.

Key workers Usually public sector workers (e.g. firemen and teachers) who the government have judged are essential to the social and economic infrastructure of areas. Governments have tried to address the shortage of key workers in certain areas of the UK by promoting affordable housing supply initiatives, notably through councils and housing associations.

Relets Properties which are available for letting again, following the previous tenant moving out, dead or alive.

National Housing Federation Elected trade body representing the interests of housing associations to the Housing Corporation, government, councils collectively, private lenders and other interest groups and stakeholders.

Thames Gateway A growth area defined by the 2003 Sustainable Communities Plan, earmarked for housing and related infrastructure development to help balance supply and demand.

Social engineering Politically inspired plans to try to influence the composition of a community in terms of social class mix, employment mix, attitude, etc., to achieve an 'ideal society'. Term used in a derogatory sense.

Spatial engineering Social engineering applied to settlement design, e.g. to keep social classes apart, or to integrate them in the same block, estate or settlement.

Tenure The way in which households legally occupy or 'hold' their property, e.g. through owner-occupation; tenancy; leasehold.

The right to buy The right of local authority and some housing association secure tenants to buy the home they occupy, at discount, subject to length of residence and use intended for the property. Introduced by the 1980 Housing Act, and modified by subsequent legislation (e.g. 1985 and 1996 Housing Acts).

Unitary authorities Councils which combine the functions of district and county councils, e.g. London boroughs.

Buy-to-let The practice of purchasing residential property to let it out, as an investment. The landlord is non-resident.

Assured shorthold tenancy The most common letting arrangement in the private sector, although it also can be used by housing associations on a restricted basis. Has a minimum length of six months, and can be ended by at least two months' notice. If notice is properly served, the tenant has no grounds to resist repossession.

Large scale voluntary transfer (LSVT) A scheme introduced under the terms of the *1986 Housing and Planning Act* and modified by the *1988 Housing Act* where councils can sell their stock to a housing association or housing company. The tenants then become assured tenants with certain legally protected rights, although those becoming tenants after transfer may not have those rights, unless the purchasing organisation grants them. Such transfers are subject to a ballot, the government-recognised form of testing resident opinion in this case.

Capital receipts Money from the sale of property and land.

Enablers Bodies which do not do something themselves (e.g. build housing themselves), but help others to do so (in the case of housing development, often with cash or land help). Most councils are now enablers of housing rather than direct providers.

Market renewal policy Policy introduced by the Blair New Labour government from 2003 onwards under the *Sustainable Communities Plan* to deal with abandonment and dereliction in some residential areas mainly in the North and Midlands of England, often involving demolition and where necessary the creation of new settlements or housing areas through the private sector or housing associations.

Void means the same as empty, in context of housing. 'Void property' = 'empty property'.

Urban Development Corporation A body set up by the government to co-ordinate the planning and execution of redevelopment in an areas after covering more than one local authority area. Managed by a ministerially-appointed board. Example: London Docklands Development Corporation.

Neighbourhood renewal areas Area earmarked for improvements in social and physical infrastructure, co-ordinated in the main by local authorities, attracting special government grant funding.

LAWN An initiative to encourage tenants and housing applicants in London – a high demand low supply area for social housing – to move to lower demand areas in the country such as Stoke-on-Trent and parts of the North. Commenced in 2002. Now part of the **Housing and Employment Mobility Scheme**, a national scheme grant funded by ODPM to deliver tenant mobility from local authority to UK-wide scale.

Infrastructure Essential support facilities and personnel required to help a community function socially and economically. Often divided by function into *social infrastructure* – e.g. health facilities, schools, housing, recreation, and the workers needed to supply the services; and *economic infrastructure* – e.g. factory buildings and equipment. Other examples of infrastructure include 'hard' items such as telecommunications linkages, roads, railways, bridges and flood defences, and 'soft' items such as fire personnel (people), and leisure and recreation clubs and networks.

Chapter 2

Contractual and statutory tenancy conditions Contractual tenancy conditions are those granted to tenants by their landlord, in addition to those guaranteed by law. Statutory conditions are landlord and tenants' rights and obligations stated in case law, and interpreted through case law.

Octavia Hill A nineteenth-century charitable housing manager who saw housing management as a way of improving the moral and social skills of tenants just as much as managing and maintaining the dwellings they live in.

Tenant Participation Compact Requirement on local authorities, to produce a strategy and signed-up document, negotiated with tenants and leaseholders, to plan for and implement the level of participation and involvement desired, within available funds and other resources.

Niche demographic communities A demographic community is a community characterised by a specific age profile e.g. elderly community, which can be expressed in spatial or (less meaningfully) category terms (for example, an elderly persons' scheme, or all households of pensionable age in a country). A niche demographic community is a community within the general one e.g. spatially, a scheme for the frail elderly.

Marginalisation The process of ensuring that the social, economic and/or political power of a community/gender group etc., remains low in proportion to numbers in the population. Can also be used to describe the outcome of after their own needs, under Part Three of the *National Assistance Act 1948*.

Rachman A slum landlord who operated in the Notting Hill area of central West London in the 1950s and who used bullying tactics illegally to evict tenants. Gave his name to a generic term for slum landlordism.

Direct labour organisations (DLOs) The sections containing staff employed directly by the landlord to do works.

Approved development programme The Housing Corporation's three-year development plan realised through housing associations and backed partly by social housing grant allocations.

Outsourcing Employing a contractor to perform some or all of the landlord's obligations, e.g. a council employing a housing association to manage and maintain its sheltered housing.

Chartered Institute of Housing Social housing's professional body, researching and suggesting housing policy directions, awarding qualifications, running training, publishing good practice guides and housing books, and running the industry's premier annual housing conference.

Chapter 3

1978 Winter of Discontent In 1978–79 public service strikes, mainly for higher wage settlements in the face of high inflation, caused the collapse of the Callaghan government and led to changes in management of the economy in general and the money supply in particular.

Mortgage interest tax relief at source Now phased out, this was a reduction in the effective interest rate payable by mortgage holders, based on tax relief on a tax on what the property could have been rented out for, which was abolished in the 1970s. Effectively, just a state subsidy to mortgage payers.

Notional income and notional expenditure The HRA income and expenditure assumptions used by the government to decide how much subsidy councils should receive, as opposed to the actual amounts.

this activity. Can be an overt or covert activity of the state, formally or informally defined, or a product of the socio-economic arrangements existing in a group of states, state, or area(s) within them.

Dysfunction A malfunction in a component of society which threatens social harmony or the 'normal' operation of that society generally or in a specific area, spatial or functionally – for example, anti-social behaviour on behalf of young people; criminality generally; substance abuse. Groups which consistently display behaviour which goes against the standards accepted in wider society and threatens to disrupt its operation are sometimes known as dysfunctional groups in sociological studies.

Keynsianism Economic philosophy developed by John Maynard Keynes in the 1930s, maintaining that public expenditure supported through taxation was a foundation of economic growth and prosperity, in creating employment and stimulating and boosting economic output. Followed by successive Labour governments between 1945 and 1978.

Welfare state Cradle-to-grave state-funded social provision to ensure that everybody receives a reasonable level of education, food, clothing and housing, mainly by providing contributory and non-contributory redistributional finance support, without risking failure to provide for themselves through the market. Developed in its modern form by William Beveridge, the Social Services Minister, of the 1945 Labour Attlee administration.

Monetarism Economic theory developed by *Milton Friedman* in the 1960s, based on the idea that money behaves like other commodities in the marketplace, with its value determined by the relativities of supply and demand (like other commodities), and the velocity of currency in circulation. According to Friedman, inflation is caused by money supply significantly exceeding demand, causing its value to reduce, and its control is essential to economic stability. Successive governments since 1979 have tried to control inflation largely through controlling the amount of money in circulation, by credit controls, interest rate policy and notably by controlling the expenditure of public bodies,

Department for Work and Pensions Department of state responsible for the social security budget, including housing benefit and income support.

Total cost indicators (TCIs) Housing Corporation estimates of what it should cost to build a home of a given size in a given region of the country, used to help determine Social Housing grant levels.

General Fund The local authority's corporate account, into which council tax, redistributed business rates and revenue support grant (central government support) is paid, and which spends on the council's services and debt liabilities.

Ring-fencing The practice of disallowing contributions from other parts of the General Fund to cross-subisidise the activity paid for from a specific account from general income.

Social Housing Grant (SHG) Grant given by the Housing Corporation to support the development activities of housing associations. The gap between total development costs and SHG has to be made up by private borrowing and/or the association's cash reserves and/or surpluses from other schemes.

Back-payment Reimbursement.

'Notional capital receipt' Basically, an extension of a council's credit, occurring through spending capital receipts on revenue activities. Effectively abolished by the 1989 Local Government and Housing Act, they allowed councils to 'back' leases which would otherwise have eaten into the Credit Approvals and would have been used to borrow against for capital works such as major improvements.

The cascade Abolished under the *1989 Act.* Councils were only allowed to spend a set percentage of their capital receipts in the year realised. The rest was either used to redeem debt, used for something else, saved, or carried forward the next financial year. If carried forward, the same limited percentage of the remaining receipt could be spent. Receipts could be carried forward like this, until completely spent up. If cascade receipts were spent on non-capital items, a Notional Capital Receipt was created.

Reserved and useable capital receipts (RCRs and UCRs) Useable capital receipts were the amount realised from sales that could be used for capital activities. In 2003/04, this was 25% of housing and 50% of housing land receipts. The remainder, the reserved capital receipt, had to be kept back for debt redemption. From 2004/05, RCRs have been collected by government, pooled nationally and redistributed to finance housing development, major repairs and improvements to social housing generally.

Major repairs allowance Part of HRA subsidy, paid to support housing expenditure on major repairs, to replace worn-out components and keep properties in the state they should be, but not to improve them.

(Housing) private finance initiative (PFI) A vehicle whereby councils can guarantee private sector loans of 30 years for developers to construct or improve housing they own, contracting with a consortium usually made up of a bank, building society or other financial institution, and a housing company or association, to deliver improvements or construction over the life of the loan.

AME (annually managed expenditure). The Treasury's spending target estimate covering volatile public current account spending, such as Social Security, set on an annual basis, unlike Departmental Expenditure Limits (DEL) which relate to capital and other plannable programmes. Together, they make up total managed expenditure – the total public sector expenditure requirement.

Transitional arrangements Government finance policy to ease resource reductions in the early years or to the date of full implementation of a 'reform'. For example, reducing management and maintenance subsidy gradually in real terms, to provide a relatively smooth incline towards much lower 'target' allowances implied by the calculation.

Rebasing Increasing part of the HRA subsidy allowance to help compensate for losses due to reform of another part. Generally paid for by slicing off some of the subsidy increases which other councils would have realised if the subsidy formula were applied strictly. Usually tails off over a period: a form of transitional protection to 'loser' councils.

Stock conditions survey A survey undertaken by a firm of chartered surveyors or similar professionals to give a picture of the physical condition of housing owned by a body at a point in time, usually accompanied by a projection of the cost of bringing the stock up to a given standard over a term of years.

'Reckonable' items of expenditure Items of HRA expenditure which are eligible for HRA subsidy support.

Hobson's choice An option which is imposed. The only 'option' available!

Clawback Grabbing back something which has been paid out – for example, central government reclaiming part of HRA Subsidy paid out.

Opportunity cost What else you could have done with the money you have spent on something.

Rent rebate Old-fashioned term for the reduction in council tenants' rents through the housing benefit rules.

Rent allowances Old-fashioned term for the amount paid to private tenants to help them pay their rent.

'Pathfinders' Organisations used to road-test a new policy before general introduction: in this context. Councils and housing associations testing the local housing allowance.

Chapter 4

Affordable housing Housing which is priced so that people on lower than average incomes who are unable to compete on the open for sale or rental market can afford to buy or rent them. There is no definitive definition of 'affordable'.

Demography The study of population trends and characteristics, e.g. birth, death and migration rates.

Turnover rate The rate at which lets from existing stock become available over a given time period (e.g. 10 relets per month). Often expressed as a percentage of total stock. Turnover is added to new lets availability to define total letting availability.

Concealed household Household living in the same property as another household which may not be counted as separate from the other because they are related, even though (arguably) they should be, for purposes of assessing housing need. For example, a couple living with one of the pair's immediate family.

Gross housing requirement Total housing requirement not taking account of availability of properties.

Income:property price ratio Relationship between household income and the price of property, either in rental or sale terms, sometimes expressed in the cost of paying a mortgage on a property acquired at a given cost. This ratio is sometimes used to try to define what is and is not affordable. For example, an income:property price ratio of 1:4 may be considered the limit of affordability, where income may include housing support, such as housing benefit.

'Sub-market' accommodation Housing available at rates below the local market price for an equivalent property, either in terms of rent or purchase price.

HM Land Registry Authoritative government-backed source of information on property prices and trends, available on a quarterly basis.

Non-linearity Statistical trend which deviates from a straight line.

Modern methods of construction An in-vogue phrase for building using cheap materials and factory prefabrication off-site to reduce the cost of the product.

Le Corbusier 1920s French architect who developed the concept of 'villages in the sky' – tower blocks including recreation, retail and other community-sustaining elements.

1946 New Towns Act Act of Parliament which enabled the building of the UK's new towns, for example Harlow, Stevenage and Hatfield.

Market renewal Policy introduced in 2002 to deal with low-demand housing areas, by encouraging people to move in to the area, regeneration, or plain bulldozing.

'Totted' (past tense of totting) Slang term for removing fixtures and fittings of value from derelict or otherwise vacant properties, e.g. fireplace surrounds, cornices, ornamental tiles, cast-iron guttering, lead from the roof, etc. Usually involves resale to enhance other properties.

Sub-region A collection of boroughs, forming a subdivision in a region, grouped together to pursue common housing development and allocation policies; e.g. the north London sub-region, which is a subset of the 33 boroughs which make up the London region.

RSL consortium A collection of RSLs which have got together to pursue common aims, for example, to develop housing together in a cost-effective manner, using the backing of a single grant and/or loan.

Net present value The value of an income or expenditure stream (or income minus expenditure) at today's value, i.e. discounted by inflation power the term of years.

'Liveability' Jargon term used as shorthand for design and layout which enables people to enjoy their homes and surroundings due to customer-responsive design, taking account of household lifecycle changes, disability and impairment, the need for recreation, education, health and shopping facilities, etc in reasonable proximity.

Land-banking The practice of buying development land and holding onto it, developing it only when demand for housing has reached a level where required profits can be made from sales. Often used to create an artificial shortage of land, thereby forcing land prices up, for resale at a profit.

Index